RAND NATIONAL DEFENSE RESEARCH INSTITUTE

Assessing the Implications of Allowing Transgender Personnel to Serve Openly

Agnes Gereben Schaefer, Radha Iyengar,

Srikanth Kadiyala, Jennifer Kavanagh, Charles C. Engel,

Kayla M. Williams, Amii M. Kress

Prepared for the Office of the Secretary of Defense

For more information on this publication, visit www.rand.org/t/RR1530

Library of Congress Cataloging-in-Publication Data is available for this publication.
ISBN: 978-0-8330-9436-0

Published by the RAND Corporation, Santa Monica, Calif.

© Copyright 2016 RAND Corporation

RAND® is a registered trademark.

Support RAND
Make a tax-deductible charitable contribution at
www.rand.org/giving/contribute

www.rand.org

Preface

U.S. Department of Defense (DoD) policies have rendered both the physical and psychological aspects of "transgender conditions" as disqualifying conditions for accession and allow for the administrative discharge of service members who fall into these categories. However, in July 2015, Secretary of Defense Ashton Carter announced that DoD would "create a working group to study the policy and readiness implications of welcoming transgender persons to serve openly." In addition, he directed that "decision authority in all administrative discharges for those diagnosed with gender dysphoria[1] or who identify themselves as transgender be elevated to the Under Secretary of Defense (Personnel and Readiness), who will make determinations on all potential separations" (DoD, 2015b).

It is against this backdrop that DoD is considering allowing transgender personnel to serve openly. To assist in identifying the potential implications of such a change in policy, the Office of the Under Secretary of Defense for Personnel and Readiness asked the RAND National Defense Research Institute to conduct a study to (1) identify the health care needs of the transgender population, transgender service members' potential health care utilization rates, and the costs associated with extending health care coverage for transition-related treatments; (2) assess the potential readiness implications of allowing transgender service members to serve openly; and (3) review the experiences of foreign militaries that permit transgender service members to serve openly. This report documents the findings from that study. This research should be of interest to DoD and military service leadership, members of Congress, and others who are interested in the potential implications of allowing transgender personnel to serve openly in the U.S. armed forces.

This research was sponsored by the Office of the Under Secretary of Defense for Personnel and Readiness and conducted within the Forces and Resources Policy Center of the RAND National Defense Research Institute, a federally funded research and development center sponsored by the Office of the Secretary of Defense, the Joint

---

[1] *Gender dysphoria* is "discomfort or distress that is caused by a discrepancy between a person's gender identity and that person's sex assigned at birth" (World Professional Association for Transgender Health, 2011, p. 2).

Staff, the Unified Combatant Commands, the Navy, the Marine Corps, the defense agencies, and the defense Intelligence Community.

For more information on the RAND Forces and Resources Policy Center, see www.rand.org/nsrd/ndri/centers/frp or contact the director (contact information is provided on the web page).

Contents

Figures and Tables

Figures

Tables

Summary

The U.S. Department of Defense (DoD) is reviewing its policy on transgender person-nel serving openly and receiving gender transition–related treatment during military service. The prospect of transgender personnel serving openly raises a number of policy questions, including those regarding access to gender transition–related health care, the range of transition-related treatments to be provided, the potential costs associated with these treatments, and the impact of gender transition–related health care needs (i.e., surgical, pharmacologic, and psychosocial) on military readiness—specifically, in terms of the deployability of transgender service members. The Office of the Under Secretary of Defense for Personnel and Readiness asked the RAND National Defense Research Institute to conduct a study to (1) identify the health care needs of the trans-gender population, transgender service members' potential health care utilization rates, and the costs associated with extending health care coverage for transition-related treatments; (2) assess the potential readiness implications of allowing transgender ser-vice members to serve openly; and (3) review the experiences of foreign militaries that permit transgender service members to serve openly. This report presents the study findings centered around the following research questions:

- What are the health care needs of the transgender population?
- What is the estimated transgender population in the U.S. military?
- How many transgender service members are likely to seek gender transition–related medical treatment?
- What are the costs associated with extending health care coverage for gender transition–related treatments?
- What are the potential readiness implications of allowing transgender service members to serve openly?
- What lessons can be learned from foreign militaries that permit transgender per-sonnel to serve openly?
- Which DoD policies would need to be changed if transgender service members are allowed to serve openly?

In the following sections, we summarize the findings associated with each research question.

What Are the Health Care Needs of the Transgender Population?

For the purposes of this analysis, we use *transgender* as an umbrella term to refer to individuals who identify with a gender different from the sex they were assigned at birth. Under the recently established criteria and terminology in the fifth edition of the *Diagnostic and Statistical Manual of Mental Disorders* (DSM-5), the American Psychiatric Association (APA) publication that provides standard language and criteria for classifying mental health conditions, transgender status alone does not constitute a medical condition (APA, 2013). Instead, under the revised diagnostic guidelines, only transgender individuals who experience significant related distress are considered to have a medical condition called *gender dysphoria* (GD). Some combination of psychosocial, pharmacologic (mainly but not exclusively hormonal), or surgical care may be medically necessary for these individuals. Psychotherapy to confirm a diagnosis of GD is a common first step in the process, often followed by hormone therapy and, perhaps, gender reassignment surgery involving secondary or primary sex characteristics. Not all individuals seek all forms of care.

A subset of transgender individuals may choose to *transition*, the term we use to refer to the act of living and working as a gender different from that assigned at birth. For some, the transition may be primarily social, with no accompanying medical treatment; we refer to this as *social transition*. For others, medical treatments, such as hormone therapy and hair removal, are important steps to align their physical body with their target gender. We refer to this as *medical transition*. A subset of those who medically transition may choose to undergo gender reassignment surgery to make their body as congruent as possible with their gender identity. This process of surgical transition is also often referred to as *sex* or *gender reassignment* or *gender confirmation*.

What Is the Estimated Transgender Population in the U.S. Military?

Estimates of the transgender population in the U.S. military and the analyses presented in this report should be interpreted with caution, as there have been no rigorous epidemiological studies of the size or health care needs of either the transgender population in the United States or the transgender population serving in the military. As a result, much existing research relies on self-reported, nonrepresentative survey samples. We applied a range of prevalence estimates from published research to fiscal year (FY) 2014 personnel numbers to estimate the number of transgender individuals serving in the U.S. military. We estimate that there are between 1,320 and

6,630 transgender personnel serving in the active component (AC) and 830–4,160 in the Selected Reserve (SR). Combining survey evidence from multiple states and adjusting for the male/female distribution in the military gave us a midrange estimate of around 2,450 transgender personnel in the AC and 1,510 in the SR.

How Many Transgender Service Members Are Likely to Seek Gender Transition–Related Medical Treatment?

We developed two estimates of demand for gender transition–related medical treatments based on private health insurance data and self-reported data from the National Transgender Discrimination Survey (NTDS). Based on our analyses of available private health insurance data on transition-related health care utilization, we expect only a small number of AC service members to access transition-related health care each year. Our estimates based on private health insurance data ranged from 0.022 to 0.0396 annual claimants per 1,000 individuals. Applied to the AC population, these estimates led to a lower-bound estimate of 29 AC service members and an upper-bound estimate of 129 AC service members annually utilizing transition-related health care, out of a total AC force of 1,326,273 in FY 2014.

We also projected health care utilization using the estimated prevalence of transgender service members and self-reported survey data from the NTDS describing the proportion of the transgender population seeking transition-related treatments by age group. Based on these calculations, we estimated, as an upper-bound, 130 total gender transition–related surgeries and 140 service members initiating transition-related hormone therapy (out of a total AC force of 1,326,273 in FY 2014). To put these numbers in perspective, an estimated 278,517 AC service members accessed mental health services in FY 2014. Hence, we expect annual gender transition–related health care to be an extremely small part of the overall health care provided to the AC population.

What Are the Costs Associated with Extending Health Care Coverage for Gender Transition–Related Treatments?

To determine the budgetary implications of gender transition–related treatment for Military Health System (MHS) health care costs, we again used data from the private health insurance system on the cost of extending coverage for this care to the transgender personnel population. We estimate that AC MHS health care costs will increase by between $2.4 million and $8.4 million annually—an amount that will have little impact on and represents an exceedingly small proportion of AC health care expenditures (approximately $6 billion in FY 2014)[1] and overall DoD health care expenditures ($49.3 billion actual expenditures for the FY 2014 Unified Medical Program; Defense Health Agency, 2015, p. 22). These estimates imply small increases in annual health care costs; results that are consistent with the low prevalence of transgender personnel and the low annual utilization estimates that we identified.

What Are the Potential Readiness Implications of Allowing Transgender Service Members to Serve Openly?

Similarly, when assessing the readiness impact of a policy change, we found that less than 0.0015 percent of the total available labor-years would be affected, based on estimated gender transition–related health care utilization rates.[2] This is because even at upper-bound estimates, less than 0.1 percent of the total force would seek transition-related care that could disrupt their ability to deploy.[3] Existing data also suggest a minimal impact on unit cohesion as a result of allowing transgender personnel to serve openly. However, we caution that these results rely on data from the general civilian population and foreign militaries, as well as previous integration experiences in the military (e.g., gays, lesbians, women), which may not hold for transgender service members.

What Lessons Can Be Learned from Foreign Militaries That Permit Transgender Personnel to Serve Openly?

There are 18 countries that allow transgender personnel to serve openly in their militaries: Australia, Austria, Belgium, Bolivia, Canada, Czech Republic, Denmark, Estonia, Finland, France, Germany, Israel, Netherlands, New Zealand, Norway, Spain, Sweden, and the United Kingdom (Polchar et al., 2014). Our analysis focused on the policies of the four countries—Australia, Canada, Israel, and the United Kingdom—with the most well-developed and publicly available policies on transgender military personnel. Several common themes emerged from our analysis of their experiences:

- The service member's gender is usually considered to have shifted to the target gender in areas such as housing, uniforms, identification cards, showers, and restrooms when a service member publicly discloses an intention to live as the target

---

[1]   AC beneficiaries make up less than 15 percent of TRICARE beneficiaries (Defense Health Agency, 2015).

[2]   We define a labor-year as the amount of work done by an individual in a year.

[3]   We note that the ability to deploy is not exactly equivalent to readiness. A service member's readiness could be measured by the ability to participate in required training and exercises, which could be affected by treatments as well. Our estimates include days of inactivity due to medical treatments, which could also apply in these settings.

gender and receives a diagnosis of gender incongruence. However, physical fitness standards typically do not fully shift until the medical transition is complete. In many cases, personnel are considered exempt from physical fitness tests during transition.

- Because the gender transition process is unique for each individual, issues related to physical standards and medical readiness are typically addressed on a case-by-case basis. This flexibility has been important in addressing the needs of transgender personnel.
- The foreign militaries we analyzed permit the use of sick leave for gender transition–related medical issues and cover some, if not all, medical or surgical treatments related to a service member's gender transition.
- In no case was there any evidence of an effect on the operational effectiveness, operational readiness, or cohesion of the force.

The case studies also suggested a number of key best practices:

- Ensure strong leadership support.
- Develop an explicit written policy on all aspects of the gender transition process.
- Provide education and training to the entire force on transgender personnel policy, but integrate this training with other diversity-related training and education.
- Develop and enforce a clear anti-harassment policy that addresses harassment aimed at transgender personnel alongside other forms of harassment.
- Make subject-matter experts and gender advisers serving within military units available to commanders seeking guidance or advice on gender identity issues.
- Identify and communicate the benefits of an inclusive and diverse workforce.

Which DoD Policies Would Need to Be Changed if Transgender Service Members Are Allowed to Serve Openly?

We reviewed 20 current accession, retention, separation, and deployment regulations across the services and the Office of the Secretary of Defense to assess the impact of changes that may be required to allow transgender individuals to serve openly. We also reviewed 16 other regulations that have been replaced by more recent regulations or that did not mention transgender personnel.[4] Based on the experiences of foreign militaries, we recommend that DoD issue clear and comprehensive policies.

---

[4] These additional policies can are listed in Appendix D of this report.

Accession Policy

We recommend that DoD review and revise the language in accession instructions to match the DSM-5 for conditions related to mental fitness, ensuring the alignment of mental health–related language and facilitating appropriate screening and review processes for disorders that may affect fitness for duty. Similarly, physical fitness standards should specify physical requirements (rather than physical conditions). Finally, physical fitness policies should clarify when the service member's target gender requirements will begin to apply.

Retention Policy

We recommend that DoD expand and enhance its guidance and directives to clarify retention standards for review during and after medical transition. For example, evidence from Canada and Australia suggests that transgender personnel may need to be held medically exempt from physical fitness testing and requirements (Canadian Armed Forces, 2012; Royal Australian Air Force, 2015). However, after completing medical transition, the service member could be required to meet the standards of the acquired gender.

Separation Policy

DoD may wish to revise the current separation process based on lessons learned from the repeal of Don't Ask, Don't Tell. The current process relies on administrative decisions outside the purview of the standard medical and physical review process. This limits the documentation and review of discharges, and it could prove burdensome if transgender-related discharges become subject to re-review and redetermination. When medically appropriate, DoD may wish to establish guidance on when such discharge reviews should be handled through the existing medical fitness processes. We also recommend that DoD develop and disseminate clear criteria for assessing whether and how transgender-related conditions may interfere with duty performance.

Deployment Policy

The degree of austerity will differ across deployment environments, and some locations may be able to meet the health care needs of some transgender individuals. Moreover, recent advancements can minimize the invasiveness of treatments and allow for telemedicine or other forms of remote medical care.

Given this, DoD may wish to adjust some of its processes and deployment restrictions in the context of medical and technological advancements (e.g., minimally invasive treatments, telemedicine). Such reforms could minimize the readiness impact of medical procedures that are common among the transgender population. For example, current regulations specifying that conditions requiring regular laboratory visits that cannot be accommodated in a deployed environment can leave service members ineligible for deployment and would affect all individuals receiving hormone treatments

(Office of the Assistant Secretary of Defense for Health Affairs, 2013, p. 3). These treatments require laboratory monitoring every three months for the first year as hormone levels stabilize (Hembree et al., 2009; Elders et al., 2014). To avoid this cost, DoD would need to either permit more flexible monitoring strategies[5] or provide training to deployed medical personnel.[6]

---

[5] Some experts suggest that alternatives, such as telehealth reviews, would address this issue for rural populations with limited access to medical care (see, for example, World Professional Association for Transgender Health, 2011).

[6] "Independent duty corpsmen, physician assistants, and nurses can supervise hormone treatment initiated by a physician" (Elders et al., 2014).

Acknowledgments

The authors would like to extend thanks to our DoD sponsors who provided valuable feedback on various briefings over the course of this study. Deputy Assistant Secretary of Defense for Military Personnel Policy Anthony Kurta was also extremely helpful in providing oversight of this research effort.

We also benefited from the contributions of our RAND colleagues. Bernard Rostker, Michael Johnson, John Winkler, Lisa Harrington, Kristie Gore, and Sarah Meadows provided helpful formal peer reviews of this report. Michelle McMullen provided administrative support, and Lauren Skrabala provided editorial assistance.

We thank them all, but we retain full responsibility for the objectivity, accuracy, and analytic integrity of the work presented here.

Abbreviations

AC	active component
APA	American Psychiatric Association
DoD	U.S. Department of Defense
DoDI	U.S. Department of Defense instruction
DSM-5	*Diagnostic and Statistical Manual of Mental Disorders*, fifth ed.
FY	fiscal year
GD	gender dysphoria
IDF	Israel Defense Forces
LGBT	lesbian, gay, bisexual, and transgender
MHS	Military Health System
MTF	military treatment facility
NTDS	National Transgender Discrimination Survey
SR	Selected Reserve
VHA	Veterans Health Administration
WPATH	World Professional Association for Transgender Health

CHAPTER ONE

Introduction

U.S. Department of Defense (DoD) policies have rendered both the physical and psychological aspects of "transgender conditions" disqualifying conditions for accession and allowed for the administrative discharge of service members who fall into these categories. However, in July 2015, Secretary of Defense Ashton Carter announced that DoD would "create a working group to study the policy and readiness implications of welcoming transgender persons to serve openly." In addition, he directed that "decision authority in all administrative discharges for those diagnosed with gender dysphoria[1] or who identify themselves as transgender be elevated to the Under Secretary of Defense (Personnel and Readiness), who will make determinations on all potential separations" (DoD, 2015b). It is against this backdrop that DoD is considering allowing transgender service members to serve openly. To assist in identifying the potential implications of such a policy change, the Office of the Under Secretary of Defense for Personnel and Readiness asked the RAND National Defense Research Institute to conduct a study to (1) identify the health care needs of the transgender population, transgender service members' potential health care utilization rates, and the costs associated with extending health care coverage for transition-related treatments; (2) assess the potential readiness impacts of allowing transgender service members to serve openly; and (3) review the experiences of foreign militaries that permit transgender service members to serve openly.

Study Approach

Our study approach centered around the following research questions:

- What are the health care needs of the transgender population?
- What is the estimated transgender population in the U.S. military?

---

[1] *Gender dysphoria*, or GD, is "discomfort or distress that is caused by a discrepancy between a person's gender identity and that person's sex assigned at birth" (World Professional Association for Transgender Health [WPATH], 2011, p. 2).

- How many transgender service members are likely to seek gender transition–related medical treatment?
- What are the costs associated with extending health care coverage for gender transition–related treatments?
- What are the potential readiness implications of allowing transgender service members to serve openly?
- What lessons can be learned from foreign militaries that permit transgender personnel to serve openly?
- Which DoD policies would need to be changed if transgender service members are allowed to serve openly?

We explain our methodological approaches in detail in each chapter of this report, but, here, we present overviews of the various methodologies that we employed. We began our analysis by defining the term *transgender* and then identifying the health care needs of the transgender population. This entailed an extensive literature review of these health care needs, along with treatment standards and medical options— particularly for those who have been diagnosed with gender dysphoria (GD).

We then undertook a review of existing data to estimate the prevalence and likely utilization rates of the transgender population in the U.S. military. Based on our estimates of the potential utilization of gender transition–related health care services, we estimated the Military Health System (MHS) costs for transgender active-component (AC) service members and reviewed the potential effects on force readiness from allowing these service members to serve openly.

We adopted two distinct but related approaches to estimating health care utilization and readiness impact. The first is what we label the *prevalence-based approach*, in which we estimated the prevalence of transgender personnel in the military and applied information on rates of gender transition and reported preferences for different medical treatments to measure utilization and the implied cost and readiness impact. This approach has the benefit of including those who may seek other forms of accommodation, even if they do not seek medical care. It also provides detailed information on the types of medical treatments likely to be sought, which can improve the accuracy of cost and readiness estimates. However, this approach suffers from a lack of rigorous evidence in terms of the rates at which transgender individuals seek treatment and instead relies on the nonscientific National Transgender Discrimination Survey (NTDS). This approach also relies on prevalence measures from only two states, Massachusetts and California, which may not be directly applicable to military populations.

Using our second approach, which we label the *utilization-based approach*, we estimated the rates of utilization of gender transition–related medical treatment. This approach has the benefit of providing real-world measures of utilization, which may be more accurate and more rigorously collected than survey information. However, it suffers from a lack of large-scale evidence and instead relies on several case studies that may not be directly applicable to the U.S. military. Given the caveats described, these approaches provide the best available estimate of the potential number of transgender service members likely to seek medical treatment or require readiness-related accommodations.[2] In both cases, we applied measures of population prevalence and utilization to fiscal year (FY) 2014 DoD force size estimates to provide estimates of prevalence within the U.S. military.

We also reviewed the policies of foreign militaries that allow transgender service members to serve openly. Our primary method supporting the observations presented in this report was an extensive document review that included primarily publicly available policy documents, research articles, and news sources that discussed policies on transgender personnel in these countries. The information about the transgender personnel policies of foreign militaries came directly from the policies of these countries, as well as from research articles describing the policies and their implementation. Findings on the effects of open transgender service on cohesion and readiness drew largely from research articles that specifically examined this question using interviews and an analysis of studies completed by the foreign militaries themselves. Finally, insights on best practices and lessons learned emerged both directly from research articles describing the evolution of policy and experience and indirectly from commonalities in the policies and experiences of our four in-depth case studies. Recommendations provided in this report are based on these best practices and lessons learned, as well as a consideration of the unique characteristics of the U.S. military.

Finally, for our analysis of DoD policies, we reviewed 20 current accession, retention, separation, and deployment regulations across the services and the Office of the Secretary of Defense. We also reviewed 16 other regulations that have been replaced by more recent regulations or that did not mention transgender personnel.[3] Our review focused on transgender-specific DoD instructions (DoDIs) that may contain unnecessarily restrictive conditions and reflect outdated terminology and assessment processes. However, in simply removing these restrictions, DoD could inadvertently affect standards overall. While we focused on reforms to specific instructions and directives, we note that DoD may wish to conduct a more expansive review of personnel policies to ensure that individuals who join and remain in service can perform at the desired level, regardless of gender identity.

Limitations and Caveats

A critical limitation of such a comprehensive assessment is the lack of rigorous epidemiological studies of the size or health care needs of either the U.S. transgender population or the transgender population serving in the military. Indeed, much of the

---

[2] We define *accommodations* as adjustments in military rules and policies to allow individuals to live and work in their target gender.

[3] These additional policies are listed in Appendix D of this report.

existing research on the transgender population relies on self-reported, nonrepresentative survey data, along with unstandardized calculations using results from available studies. Because there are no definitive data on this topic, the information presented here should be interpreted with caution and, therefore, we present the full range of estimates.

Organization of This Report

The report is organized around our seven research questions. Chapter Two defines what is meant by the term *transgender*, identifies the health care needs of the transgender population, explains the various treatment options for those diagnosed with GD, and examines the capacity of the MHS to provide treatment options to service members diagnosed with GD. Chapter Three estimates the number of transgender service members in the AC and Selected Reserve (SR). Chapter Four estimates how many transgender service members are likely to seek medical treatment. Chapter Five estimates the costs associated with extending health care coverage for gender transition–related treatments. Chapter Six assesses the potential readiness implications of allowing transgender service members to serve openly. Chapter Seven identifies lessons learned from foreign militaries that allow transgender personnel to serve openly. Chapter Eight offers recommendations regarding which DoD accession, retention, separation, and deployment policies would need to be changed if a decision is made to allow transgender service members to serve openly. Chapter Nine summarizes key findings presented in the report and suggests best practices for implementing policy changes.

Appendix A presents definitions of common terms related to gender transition and transgender identity. Appendix B provides a history of the historical nomenclature associated with transgender identity. Appendix C provides details on the psychosocial, pharmacologic, surgical, and other treatments for GD. Appendix D lists the DoD accession, retention, separation, and deployment policies that we reviewed.

CHAPTER TWO

What Are the Health Care Needs of the Transgender Population?

This report begins by describing the health care needs of the U.S. transgender population overall. To discern the potential impact of changing DoD policies to allow transgender military personnel to serve openly and to ensure appropriate health care for those who seek gender transition–related treatment, it is also important to consider whether the MHS has the capacity to provide this care.

Definitions of Key Terms and Concepts

A challenge to our efforts to understand the health care needs of the transgender population in general, as well as in the military, is the varied and shifting terminology used in the clinical literature. Consequently, here, we define a range of terms that we will use throughout this review.[1] Consistent with the fifth edition of the *Diagnostic and Statistical Manual of Mental Disorders* (DSM-5), the American Psychiatric Association (APA) publication that provides standard language and criteria for classifying mental health conditions, we use the term *transgender* to refer to "the broad spectrum of individuals who . . . identify with a gender different from their natal gender" (APA, 2013).[2] *Natal gender* or *birth sex*, which is the sex that an individual was assigned at birth and typically correlates with primary sex characteristics (e.g., genitalia).

We refer to the subset of the population whose gender identity does not conform with the expressions and behaviors typically associated with the sex to which they were assigned at birth as *transgender* or *gender nonconforming*. Many identities fall under these umbrella terms, including individuals who identify as androgynous, multigendered, third gender, and two-spirit people. The *gender nonconforming* category also includes individuals who *cross-dress*, which means they wear clothing that is traditionally worn by a gender different from that of their birth sex. The exact definitions of each of these identities vary under the term *gender nonconforming*, and individuals may

---

[1]   A comprehensive list of terms and definitions is provided in Appendix A.

[2]   A brief history of the DSM language and diagnostic criteria for related conditions is presented in Appendix B.

fluidly change, blend, or alter their gender identity over time. For the purposes of this analysis, we use *transgender* as an umbrella term that refers to individuals who identify with a gender different from the sex they were assigned at birth.

Importantly, under the recently established criteria and terminology outlined in DSM-5, transgender status alone does not constitute a medical condition (APA, 2013). Instead, under the revised diagnostic guidelines, only transgender individuals who experience significant related distress are considered to have a medical condition called *gender dysphoria* (GD). Some combination of psychosocial, pharmacologic (mainly but not exclusively hormonal), or surgical care may be medically necessary for these individuals. Psychotherapy to confirm a diagnosis of GD is a common first step in the process, often followed by hormone therapy and, perhaps, by gender reassignment surgery involving secondary or primary sex characteristics. Not all patients seek all forms of care. However, recognized standards of care require documentation of 12 continuous months of hormone therapy and living in the target gender role consistently and in all aspects of life. Unfortunately, the diagnosis is newly established, and data from which to estimate the size of these subgroups are lacking. In the future, however, transgender individuals seeking gender transition–related treatment are likely to require a GD diagnosis as the clinical justification.

Among transgender individuals, a subset may choose to *transition*, the term used to refer to the act of living and working in a gender different from one's sex assigned at birth. For some individuals, this may involve primarily social change but no medical treatment; this is referred to as *social transition*. For others, medical treatments, such as hormone therapy and hair removal, are important steps to align their physical body with their target gender. This is referred to as *medical transition*. A subset of those who medically transition may choose to undergo *gender reassignment surgery* to make their physical body as congruent as possible with their gender identity. This process of *surgical transition* is also often referred to as *sex* or *gender reassignment* or *gender confirmation*.

Health Care Needs of the Transgender Population

The main types of gender transition–related treatments are psychosocial, pharmacologic (primarily but not exclusively hormonal), and surgical. While one or more of these types of treatments may be necessary for some transgender individuals with GD, the course of treatments varies and must be determined on an individual basis by patients and clinicians. Since little is known about currently serving transgender service members, the following discussion draws primarily from available research on nonmilitary transgender populations.[3]

---

[3]   The 2015 DoD Health Related Behavior Survey of active-duty service members was being fielded concurrently to this research. It marked the first time a U.S. military survey asked questions relating to gender identity.

Diagnosis and Treatments for Gender Dysphoria

Treatments deemed necessary for transgender populations have shifted over time based on research advancements and the accumulation of clinical knowledge. The World Professional Association for Transgender Health (WPATH) regularly publishes revised versions of its *Standards of Care for the Health of Transsexual, Transgender, and Gender Nonconforming People*; the most current at the time of our research was version 7. The standards are designed to guide the treatment of patients experiencing GD while recognizing that not all expressions of gender nonconformity require treatment (WPATH, 2011, p. 2). Some transgender individuals (again, the proportion is largely unknown) experience significant dysphoria (distress) with the sex and gender they were assigned at birth, and they meet formal DSM-5 diagnostic criteria for GD, as described in Appendix B of this report. For those diagnosed with GD, treatment options include psychotherapy, hormone therapy, surgery, and changes to gender expression and role (i.e., how people present themselves to the world; WPATH, 2011, pp. 9–10). We discuss these treatment options in detail in Appendix C.

Not all patients will prefer or need all or any of these options; however, when clinically indicated, appropriate care can "alleviate gender dysphoria by bringing one's physical characteristics into alignment with one's internal sense of gender" (Herman, 2013b, p. 4). There have been no randomized controlled trials of the effectiveness of various forms of treatment, and most evidence comes from retrospective studies. The widely endorsed consensus-based practice guidelines outlined in the WPATH *Standards of Care* suggest that transition-related mental health care, hormone therapy, and surgery are generally effective and constitute necessary health care for many individuals with GD.[4] The appropriate treatment plan is best determined collaboratively by patients and their health care providers. Optimally, specialized transgender health care will be provided by an interdisciplinary team (WPATH, 2011, p. 26).

Military Health System Capacity and Gender Transition–Related Treatment

To discern the potential impact of changing DoD policies to allow transgender military personnel to serve openly and to ensure appropriate health care for GD, it is also important to consider whether the MHS has the capacity to provide this care.

We anticipate that these survey results will provide additional information regarding how many transgender personnel currently serve in the U.S. military and their health behaviors.

[4]   These standards are endorsed by the American Medical Association, American Psychological Association, American Academy of Family Physicians, National Association of Social Workers, World Professional Association for Transgender Health, and American College of Obstetricians and Gynecologists (see Lambda Legal, 2012). Major insurers, including Aetna and UnitedHealthcare, have incorporated many of these standards of care into their policies (see, for example UnitedHealthcare, 2015).

Psychotherapy, Hormone Therapies, and Gender Transition–Related Surgery

Both psychotherapy and hormone therapies are available and regularly provided through the military's direct care system, though providers would need some additional continuing education to develop clinical and cultural competence for the proper care of transgender patients. Surgical procedures quite similar to those used for gender transition are already performed within the MHS for other clinical indications.

Reconstructive Surgery

Reconstructive breast/chest and genital surgeries are currently performed on patients who have had cancer, been in vehicular and other accidents, or been wounded in combat. The skills and competencies required to perform these procedures on transgender patients are often identical or overlapping. For instance, mastectomies are the same for breast cancer patients and female-to-male transgender patients. Perhaps most importantly, the surgical skills and competencies for some gender transition surgeries also overlap with skills required for the repair of genital injuries sustained in combat, which have increased dramatically among troops deployed to Afghanistan. From 2009 to 2010, the percentage of wounded troops with genitourinary injuries transiting through Landstuhl Regional Medical Center in Germany nearly doubled from 4.8 percent to 9.1 percent—a dramatic increase that led some health providers to call this the "new 'signature wound'" of Operation Enduring Freedom (D. Brown, 2011).[5] There are particular similarities to the procedures recommended to treat those experiencing dismounted complex blast injuries, which typically involve multiple amputations with other injuries, often to the genitals (Wallace, 2012). Providing high-quality surgery to treat the 5 percent of combat wounds that require penile reconstruction requires extensive knowledge and practice in reconstructive techniques (Williams and Jezior, 2013). Assuming the MHS continues to directly provide health services as it has in the past, there are at least two potential implications: First, military surgeons may currently have the competencies required to surgically treat patients with GD, and, second, performing these surgeries on transgender patients may help maintain a vitally important skill required of military surgeons to effectively treat combat injuries during a period in which fewer combat injuries are sustained.

Cosmetic Surgery

Recognition of the requirement for reconstructive plastic surgery as a result of the wartime mission drives the existing DoD policy for cosmetic surgery procedures in the MHS; the services have requirements and manpower authorizations for specialists who can perform reconstructive plastic surgery (Office of the Assistant Secretary of Defense

---

[5]   Experimental penis transplants, expected to be performed for the first time within the next year at Johns Hopkins School of Medicine, are being developed in the United States specifically for combat-wounded veterans; however, there may be benefits for transgender patients as well (Welsh, 2015).

for Health Affairs, 2005, p. 1). Cosmetic/reconstructive surgery skills need to be maintained with practice, and surgeons must also "meet board certification, recertification, and graduate medical education program requirements" (Office of the Assistant Secretary of Defense for Health Affairs, 2005, p. 1).

Current DoD policy draws a distinction between elective cosmetic plastic surgery performed "to improve the patient's appearance or self-esteem" and reconstructive plastic surgery performed on bodily structures that are abnormal due to health conditions to improve function or approximate a normal appearance (Office of the Assistant Secretary of Defense for Health Affairs, 2005, p. 3). While reconstructive surgeries constitute necessary treatment, access to elective cosmetic surgical procedures is subject to added constraints. For example, cosmetic procedures are performed on a space-available basis and restricted to those who will be TRICARE-eligible for at least six months. These procedures also require written permission from the commander of the service member's active-duty unit, and the patient must pay surgical, institutional, and anesthesia fees (Office of the Assistant Secretary of Defense for Health Affairs, 2005, p. 3).[6] DoD recognizes the need for these reconstructive surgery competencies and has crafted a policy to cover plastic surgeries to maintain providers' surgical skills and certification requirements.

Potential Consequences of Not Providing Necessary Gender Transition–Related Care

The discussion of the health care needs of transgender military personnel is incomplete without considering the potential unintended effects of constraining or limiting gender transition–related treatment. Little question remains that there are transgender personnel currently serving in the AC. Adverse consequences of not providing transition-related health care to transgender personnel could include avoidance of other necessary health care, such as important preventive services, as well as increased rates of mental and substance use disorders, suicide, and reduced productivity.

Research indicates that, "due to discrimination and problematic interactions with health care providers, transgender individuals frequently do not access health care, resulting in short and long-term adverse health outcomes" (Roller, Sedlak, and Draucker, 2015, p. 418).[7] Further, patients denied appropriate health care may turn to other solutions, such as injecting construction-grade silicone into their bodies to alter

---

[6] Interestingly, according to Elders et al. (2014, p. 19), there is no difference in leave policies related to recovery time between the two.

[7] For example, among NTDS respondents, 28 percent reported postponing or avoiding treatment when sick or injured, and 33 percent delayed or skipped preventive care due to discrimination or disrespect from health care providers (Grant et al., 2011, p. 76). In one study, transgender respondents had fewer self-reports of good health and were more likely to report limitations on daily activities due to health issues (Kates et al., 2015, p. 5).

their shape (State of California, 2012, p. 12). There are also potential costs related to mental health care services for individuals who do not receive such care (Herman, 2013b, p. 20). Multiple observational studies have suggested significant and sometimes dramatic reductions in suicidality, suicide attempts, and suicides among transgender patients after receiving transition-related treatment (State of California, 2012, p. 10). A study by Padula, Heru, and Campbell (2015) found that removing exclusions on transgender care "could change the trajectory of health for all transgender persons" at a minimal cost per member per month.[8]

However, we caution that it is not known how well these findings generalize to military personnel. Moreover, while the existing data offer some indication of the needs for and costs of gender transition–related health care, it is important to note that none of these studies were randomized controlled trials (the gold standard for determining treatment efficacy). In the absence of quality randomized trial evidence, it is difficult to fully assess the outcomes of treatment for GD.

---

[8]   Specifically, they found that insurance provider coverage for transgender-related services resulted in "greater effectiveness, and was cost-effective relative to no health benefits at 5 and 10 years from a willingness-to-pay threshold of $100,000/[quality-adjusted life year]."

CHAPTER THREE

What Is the Estimated Transgender Population in the U.S. Military?

This chapter provides several estimates of the number of transgender service members in the U.S. military. To date, there have been no systematic studies of the number of transgender individuals in the U.S. general population or in the U.S. military. Current studies rely on clinical samples of health care service utilizers, nonrepresentative samples assembled in ways that are difficult to replicate, and self-reported survey data from a small number of states.

General Population Estimates of Transgender Prevalence

The transgender prevalence in the U.S. general population is thought to be significantly less than 1 percent (Gates, 2011, p. 6; APA, 2013, p. 454). However, there have been no rigorous epidemiological studies in the general U.S. population that confirm this estimate. Our subsequent estimates must be qualified, therefore, as somewhat speculative; they are based on numerous sources, including health services claims data, representative state-level health surveillance survey data, a convenience (i.e., nonrepresentative) sample recruited by an advocacy network, the experiences of foreign militaries, and selected other data sources.

The Williams Institute at the University of California, Los Angeles, School of Law, calculated that, based on estimates from Massachusetts and California, 0.3 percent of the U.S. population is transgender (Gates, 2011, p. 6). The Massachusetts data were collected between 2007 and 2009 as part of the Massachusetts Behavioral Risk Factor Surveillance System initiative. The survey suggests that 0.5 percent of the population in Massachusetts identifies as "transgender" (95-percent confidence interval: 0.3 to 0.6 percent; Conron et al., 2012). The California data combine information on the percentage of individuals who are transgender from the California Lesbian, Gay, Bisexual, and Transgender (LGBT) Tobacco Survey and the percentage of the overall population that is LGBT from the 2009 California Health Interview Survey. Gates multiplies these values together to estimate that 0.1 percent of the population of California is transgender.[1]

To develop an estimate of transgender prevalence for the entire United States, Gates (2011) simply averages the Massachusetts and California values, yielding 0.25 percent, then rounds that up to 0.3 percent. This measure is very problematic, however. While survey-based estimates of transgender prevalence are likely to be accurate measures of true state-level transgender prevalence, it is not clear that taking an unweighted average from states with vastly different population sizes is appropriate for estimating national prevalence. For example, a weighted average calculation using the 2009 census population estimates for California and Massachusetts implies a 0.16 percent "national" prevalence estimate, as opposed to the 0.3 percent estimate calculated by Gates (2011)—a nearly 50-percent difference. We used this 0.16 percent weighted average as our combined, national estimate using the California and Massachusetts studies. This estimate was our midrange starting point, though we included both the 0.1 percent (from California) and 0.5 percent (from Massachusetts) as comparison points.

We note that there have been and continue to be other efforts to measure the prevalence of transgender identity in the general population. The two most prominent examples are the meta-analysis conducted by WPATH and a recent effort from the U.S. census. We did not use these estimates due to concerns that they systematically undercounted the prevalence of transgender identity for a variety of reasons detailed in the discussions that follow.

Separately, in 2007, the WPATH reviewed ten studies of prevalence with estimates for transgender individuals presenting for gender transition–related care, ranging from 1:11,900 to 1:45,000 for male-to-female transitions and 1:30,400 to 1:200,000 for female-to-male transitions (WPATH, 2011).[2] The studies cited were largely based on clinical usage. The WPATH authors note that these numbers should be considered "minimum estimates at best":

> The published figures are mostly derived from clinics where patients met criteria for severe gender dysphoria and had access to health care at those clinics. These estimates do not take into account that treatments offered in a particular clinic setting might not be perceived as affordable, useful, or acceptable by all self-identified gender dysphoric individuals in a given area. By counting only those people who

---

[1] Although Gates (2011) states that 3.2 percent of the LGBT population is transgender, we note that an earlier document (California Department of Health Services, 2004) reporting analyses from the same survey states that 2 percent of this population is transgender. We were not able to obtain the raw data and could not verify which of the two values is correct. We used the 3.2-percent estimate to calculate the California transgender prevalence estimate.

[2] The studies were Wålinder, 1968; Wålinder, 1971; Hoenig and Kenna, 1974; Eklund, Gooren, and Bezemer, 1988; Tsoi, 1988; Bakker et al., 1993; van Kesteren, Gooren, and Megens, 1996; Weitze and Osburg, 1996; De Cuypere et al., 2007; and Zucker and Lawrence, 2009.

present at clinics for a specific type of treatment, an unspecified number of gender dysphoric individuals are overlooked. (WPATH, 2011, p. 7)

Additionally, the information is based on utilization rates from the ten studies, mostly conducted in European countries, such as the United Kingdom, the Netherlands, Sweden, Germany, and Belgium. One study was conducted in Singapore. This raises concerns about the applicability of these estimates to the U.S. population due to differences in costs and social tolerance, both of which would likely make health utilization behavior in Europe significantly different from that in the United States. Moreover, the studies were conducted over a 30-year period in which utilization was dramatically increasing, suggesting that the estimates were not stable. This concern is reported in the WPATH report, with the authors noting that the trend (over time) was due to higher rates of individuals seeking care. In one example, the estimated transgender population doubled in just five years in the United Kingdom. If the numbers are increasing over time based on the use of clinics, then an estimate from ten to 15 years ago would likely be very low relative to utilization in those same places today, and again not representative of likely utilization in the United States.[3]

Harris (2015) used information on name and sex changes in Social Security Administration data files to estimate the number of transgender individuals in the U.S. population. Using information on male-to-female and female-to-male name changes, he estimates that there were 89,667 transgender individuals in the United States in 2010. Of this group, 21,833 (24 percent) also changed their sex, according to Social Security records; during some periods in U.S. history, this required documented proof of either initiation or completion of medical transition. Since name changes are not required, prevalence estimated in this manner is likely to be a lower-bound estimate of the true transgender prevalence rate in the United States. Using the 2010 population of adults age 18 and over as the denominator (234,564,071), 89,667 transgender cases implies a lower-bound transgender prevalence rate of 0.038 percent in the United States.

---

[3]   According to the WPATH authors,

The trend appears to be towards higher prevalence rates in the more recent studies, possibly indicating increasing numbers of people seeking clinical care. Support for this interpretation comes from research by Reed and colleagues (2009), who reported a doubling of the numbers of people accessing care at gender clinics in the United Kingdom every five or six years. Similarly, Zucker and colleagues (2008) reported a four- to five-fold increase in child and adolescent referrals to their Toronto, Canada clinic over a 30-year period. (WPATH, 2011, p. 7)

Prevalence-Based Approach to Estimating the Number of Transgender Service Members in the U.S. Military

Before discussing estimates of prevalence of transgender individuals in the U.S. military, it is important to note that, to our knowledge, no studies have directly measured the prevalence or incidence of transgender individuals currently serving in the active or reserve component.[4] To estimate prevalence in the military, we have constructed estimates using a combination of data sources.[5] One of those sources, the NTDS, provides detailed information on the choices and preferences of transgender individuals but it is not a randomized, representative sample of the military and thus is not generalizable.

We applied measures of population prevalence to DoD force size estimates to estimate prevalence in the U.S. military. We measured force size using information from DoD's 2014 demographics report (DoD, 2014; see Table 3.1). The demographics are separated into AC and SR. For much of the discussion of our medical care analysis, we focus on the AC. We did not include reserve-component service members, retirees, or dependents in the cost analyses because we did not have information on age and sex distribution within these beneficiary categories. Some of these beneficiary categories also have limited eligibility for health care provided through military treatment facilities (MTFs) and may receive their health care through TRICARE coverage in the purchased care setting or through other health insurance plans. For our readiness analysis, we included both the AC and SR because both components may be used for deployments. Although there are ongoing discussions regarding the feasibility of activating the Individual Ready Reserve, we excluded this population because we lacked the detailed information on gender and age needed to conduct our analysis.

Table 3.2 contains estimates of the number of transgender personnel in the AC and SR using the baseline prevalence from existing studies and shows the results of several tests that provide a range of estimates based on different assumptions in the literature. To estimate prevalence in the military, we conducted analyses using five values: (1) a lower-bound estimate of 0.1 percent based on a study in California

---

[4]   G. Brown (1988) found that eight out of 11 evaluated natal males with severe GD had a military background; he explains his findings by positing a "hypermasculine" phase among transgender individuals that coincides with the age of enlistment. Since the sample size in that study was extremely small, we do not consider this good evidence for this theory. Gates and Herman (2014) used estimates from the NTDS, combined with estimates of transgender prevalence (0.3 percent) from Gates (2011) and history of military service in the U.S. population from the American Community Survey, to estimate transgender prevalence in the military. Data from the National College of Health Administration showed that military experience was significantly higher among transgender individuals than among those who did not identify as transgender (9.4 percent versus 2.1 percent; Blosnich, Gordon, and Fine, 2015). However, these data were collected from only 51 institutions, and the response rate for the survey was only 20 percent, which again raises questions regarding the validity of the estimates.

[5]   Our estimates were constructed using Gates (2011), which combined estimates from the Massachusetts Behavioral Risk Factor Social Surveys with the California LGBT Tobacco Survey, and Gates and Herman (2014), which used data from the NTDS, Gates (2011), and the American Community Survey.

Table 3.1
DoD Military Force Demographics

Category	Number	%
Active Component		
Sex		
Female	200,692	15
Male	1,125,581	85
Age		
<25	572,293	43
26–30	293,698	22
31–35	201,137	15
36–40	137,653	11
41+	121,492	9
Total	1,326,273	—
Selected Reserve		
Sex		
Female	149,759	18
Male	682,233	82
Age		
<25	285,494	34
26–30	156,983	19
31–35	124,179	15
36–40	86,151	10
41+	179,185	22
Total	831,992	—

SOURCE: DoD, 2014.

(Conron, 2012); (2) an upper-bound estimate of 0.5 percent based on a study in Massachusetts (Gates, 2011); (3) a population-weighted average of the California and Massachusetts studies, yielding a prevalence estimate of 0.16 percent; (4) an adjustment of this population-weighted approach based on the natal male/female distribution in the military, yielding a prevalence estimate of 0.19 percent; and (5) a doubling of the population-weighted, gender-adjusted value, yielding a prevalence estimate of 0.37 percent.

Table 3.2
Prevalence-Based Estimates of the Number of Transgender Active-Component and Selected Reserve Service Members

Component	Total Force Size (FY 2014)	0.1%[a] (CA study)	0.16%[b] (combined, population-weighted CA + MA studies)	0.19%[c] (gender-adjusted rate)	0.37%[d] (twice gender-adjusted rate)	0.5%[e] (MA study)
Active	1,326,273	1,320	2,120	2,450	4,900	6,630
Selected Reserve	831,992	830	1,330	1,510	2,930	4,160

SOURCES: Estimates for force size are based on RAND calculations using FY 2014 data from DoD, 2014.

[a] Based on estimates of prevalence from a California study (Conron, 2012).

[b] Based on weighted average of studies from California and Massachusetts, weighted by relative population sizes in each state.

[c] Based on weighted average of studies from California and Massachusetts, weighted by relative population sizes in each state and applied specifically to the male/female distribution in the military components.

[d] Based on estimates of prevalence from NTDS, Gates (2011), and the American Community Survey (Gates and Herman, 2014) and applied specifically to the male/female distribution in the military.

[e] Based on estimates of prevalence from a Massachusetts study (Gates, 2011).

Based on the 0.1 percent lower bound, we estimate that there are approximately 1,320 transgender individuals in the AC and approximately 830 in the SR. Using the Massachusetts study (0.5 percent) as an upper bound, we estimate that there are approximately 6,630 transgender service members in the AC and 4,160 in the SR. Because these estimates are based on selected populations in the state and the variation in these populations is significant, we were concerned that they were not representative of broader national numbers, especially as they pertain to the gender mix of the military. Therefore, we adjusted the population-weighted combination of these estimates to account for the male/female distribution in the U.S. military populations. This gender adjustment is critical, as most research indicates that male-to-female transitions are two to three times more common than female-to-male transitions (APA, 2013; Horton, 2008; Gates, 2011; Grant et al., 2011). This assumption of a two to one difference in underlying prevalence across genders applied to the 0.16 percent aggregate estimate implies a natal male-specific prevalence of 0.2 percent and a natal female-specific prevalence of 0.1 percent. Assigning these values to the male/female AC distributions increases the military prevalence estimate from 0.16 percent to 0.19 percent, which implies that there are 2,450 transgender individuals in the AC and 1,510 in the SR.

The estimate of 0.37 percent doubles the gender-adjusted rate based on information provided by the NTDS that 20 percent of the transgender population in its sample reported a history of military service, which is twice the rate of the general population, as reported in the American Community Survey (Grant et al., 2011). We note that this is likely to be an overestimate of the overall transgender population for two reasons. First, given the highly tolerant environment in Massachusetts and California, the prevalence estimates in those two states are likely to overstate the nationwide prevalence.[6] Second, the evidence that transgender individuals are twice as likely to serve in the military is based on extrapolations from a nonrepresentative sample of individuals and not on direct, rigorous study of the transgender military population.

---

[6] For example, both California and Massachusetts are rated as "top places for LGBT rights" (Keen, 2015).

CHAPTER FOUR

How Many Transgender Service Members Are Likely to Seek Gender Transition–Related Medical Treatment?

We adopted two distinct but related approaches to estimate the health care utilization and impact on readiness of allowing transgender personnel to serve openly in the U.S. military. The first is what we label the *prevalence-based approach*, in which we estimated the prevalence of transgender individuals in the military and applied information on rates of gender transition and reported preferences for different medical treatments to measure utilization and the implied cost and readiness impact. This approach has the benefit of including those who may seek other forms of accommodation, even if they do not seek medical care. It also provides detailed information on the types of medical treatments likely to be sought, which can improve the accuracy of cost and readiness estimates. However, this approach suffers from a lack of rigorous evidence in terms of the rates at which transgender individuals seek treatment and instead relies on the nonscientific NTDS. It also relies on prevalence measures from only two states— Massachusetts and California—that may not be directly applicable to military populations.

We refer to our second approach as the *utilization-based approach*, which we used to estimate the rates of utilization of medical treatment. This approach has the benefit of providing real-world measures of utilization based on health insurance claims, which may be more accurate and more rigorously collected than survey information. However, this approach suffers from a lack of large-scale evidence and instead relies on several case studies that may not be directly applicable to the U.S. military. Despite these caveats, these approaches provide the best available estimate of the range in the potential number of transgender service members likely to seek medical treatment or require readiness-related accommodations.[1]

In both cases, we applied measures of population prevalence and utilization to DoD force size demographics to provide estimates of prevalence within the U.S. military. As indicated in the previous chapter, our calculations of population prevalence and health care utilization used FY 2014 data from DoD's 2014 demographics report (DoD, 2014; see Table 3.1 in Chapter Three).

---

[1] Again, we define *accommodations* as adjustments in military rules and policies to allow individuals to live and work in their target gender.

Prevalence-Based Approach to Estimating the Number of Gender Transition–Related Treatments in the U.S. Military

To estimate the utilization of gender transition–related health care treatments, we scaled the prevalence of transgender service members identified in Chapter Three by the rates of transition and reported take-up of medical treatments. We based our transition rates on self-reported transitions in the NTDS data. According to the NTDS, 55 percent of transgender individuals reported living and working as their target gender; we refer to this as *social transition*.[2] For others, medical treatments, such as hormone therapy and hair removal, are important steps to align their physical body with their target gender. We refer to this as *medical* or *surgical transition*. [3]

Using the prevalence estimates from Table 3.2 in Chapter Three, we used information from the NTDS on the age of transition for individuals under 25, 26–30, 31–35, 36–40, and over 40 and calibrated our estimates with the age distribution in the military. Fifty-five percent of NTDS respondents reported that they had socially transitioned over their lifetime, and the data indicate that male-to-female transition ages differ from female-to-male transition ages. Nearly 54 percent of female-to-male transitions occurred before the age of 25, compared with only 23 percent of male-to-female transitions.

We focus on social transition because we assess this as most relevant for individuals who may need accommodations as they live and work in a different gender. This was also used as the basis in some foreign militaries, as discussed in Chapter Seven. Table 4.1 presents the estimated number of individuals who may seek to transition each year under each of our prevalence assumptions. We found that a lower bound of 40 AC and 20 SR service members and an upper bound of 190 AC and 110 SR service members will seek to transition each year and may need some sort of accommodations. The population-weighted, gender-adjusted estimate implies a middle range of 65 AC and 40 SR service members who will seek to transition each year.

Next, we combine the estimates of the number of transgender service members with information on the proportion undergoing transition and the age-specific proportion undergoing gender transition–related treatment to generate the number of annual treatments. Surgical preference rates vary by transition type (male-to-female versus female-to-male transition; see Table 4.2). Surgeries are distributed evenly across

---

[2] We note that an additional 27 percent of those who had not yet socially transitioned wished to transition at some point in the future. Because the timeline and desire for transition are difficult to translate to concrete numbers, we used the estimate of 55 percent of transgender individuals living and working full-time as their target gender as our planning parameter for readiness accommodations.

[3] In the NTDS sample, 65 percent of transgender individuals had medically transitioned, and 33 percent had surgically transitioned. Note that the rate of medical transitions is higher than the rate of social transitions because some individuals receive hormone treatments but do not live full-time as their target gender.

Table 4.1
Estimated Number of Transgender Service Members Who May Seek to Transition per Year

Estimate Source	Active Component (total force: 1,326,273)	Selected Reserve (total force: 831,992)
0.1% (CA study)[a]	40	20
0.16% (combined, population-weighted CA + MA studies)[b]	60	30
0.19% (gender-adjusted rate)[c]	65	40
0.37% (twice gender-adjusted rate)[d]	130	80
0.5% (MA study)[e]	190	110

SOURCES: Estimated proportions of subgroups based on Grant et al., 2011, p. 25. Estimates for the AC and SR are based on RAND calculations using FY 2014 data from DoD, 2014.

[a] Based on estimates of prevalence from a California study (Conron, 2012).

[b] Based on weighted average of studies from California and Massachusetts, weighted by relative population sizes in each state.

[c] Based on weighted average of studies from California and Massachusetts, weighted by relative population sizes in each state and applied specifically to the male/female distribution in the military components.

[d] Based on estimates of prevalence from NTDS, Gates (2011), and the American Community Survey (Gates and Herman, 2014) and applied specifically to the male/female distribution in the military.

[e] Based on estimates of prevalence from a Massachusetts study (Gates, 2011).

NOTE: The table excludes Individual and Inactive Ready Reserve members because comparable information on their demographics was not available for analysis.

four procedures for male-to-female transitions and primarily over two procedures for female-to-male transitions.

Recall, not all of the individuals seeking to transition would meet the diagnostic criteria for GD, which is a requirement for these surgeries. Moreover, even among individuals who transition in some manner, surgical treatment rates are typically only around 20 percent, with the exception of chest surgery among female-to-male transgender individuals (see Table 4.2).

Table 4.3 shows the estimated annual number of hormone therapy treatments and surgeries in the AC and SR calculated using the same prevalence assumptions described in Chapter Three (see Table 3.2). The surgeries included in the calculations are vaginoplasty, chest surgeries, orchiectomy, hysterectomy, metoidioplasty, and phalloplasty. Note that these estimates constitute the number of treatments, not necessarily the number of individuals. For hormone therapy recipients, the number of treatments and recipients is the same, and these estimates can be treated as counts of individuals. However, the number of individuals is likely smaller for surgical counts because the

Table 4.2
Lifetime Surgery Preferences Among NTDS Survey Respondents

Procedure	Have Had (%)	Want Someday (%)	Do Not Want (%)
Male-to-female			
Augmentation mammoplasty	21	53	26
Orchiectomy	25	61	14
Vaginoplasty	23	64	14
Facial surgery	17	Not reported	Not reported
Female-to-male			
Chest surgery	43	50	7
Hysterectomy	21	58	21
Metoidioplasty	4	53	44
Phalloplasty	2	27	72

SOURCE: NTDS data (Grant et al., 2011).

NOTE: These estimates are from cross-sectional data; individuals likely received each treatment only once and varied in the age at treatment initiation.

same individual may receive more than one type of surgical treatment.[4] Using the lower-bound estimate from the California study and the upper-bound estimate from the Massachusetts study (see Table 4.3), we estimated that there will be between 45 and 220 hormone treatments and between 40 and 200 transition-related surgeries annually in the AC and SR. The combined population-weighted and gender-adjusted estimate indicates a midrange of 80 hormone treatments and 70 transition-related surgical treatments annually. Although surgical procedures are most likely to be one-time events, hormone therapy treatment rates are likely to be used indefinitely, and the cost and manpower effects will apply until individuals leave the MHS. We did not have information on the length of service conditional on age and therefore could not calculate the total number of service members who would be receiving hormone therapy at any given point in time. We recommend that this line of analysis be explored in the future.

Utilization-Based Approach to Estimating the Number of Gender Transition–Related Treatments in the U.S. Military

While the prevalence-based approach provides a tractable means to estimate potential utilization of gender transition–related care, there are a number of concerns regard-

---

[4]   For example, a female-to-male transition might include both chest surgery and phalloplasty.

Table 4.3
Estimated Annual Number of Surgeries and Hormone Therapy Users

Assumption Regarding Underlying Prevalence	Active Component		Selected Reserve	
	Annual Major Surgeries	Annual Hormone Therapy	Annual Major Surgeries	Annual Hormone Therapy
0.1% (CA study)[a]	25	30	15	15
0.16% (combined, population-weighted CA + MA studies)[b]	40	45	20	25
0.19% (gender-adjusted)[c]	45	50	25	30
0.37% (twice gender-adjusted rate)[d]	90	100	50	55
0.5% (MA study)[e]	130	140	70	80

SOURCE: RAND analysis.

[a] Based on estimates of prevalence from a California study (Conron, 2012).

[b] Based on weighted average of studies from California and Massachusetts, weighted by relative population sizes in each state.

[c] Based on weighted average of studies from California and Massachusetts, weighted by relative population sizes in each state and applied specifically to the male/female distribution in the military components.

[d] Based on estimates of prevalence from NTDS, Gates (2011), and the American Community Survey (Gates and Herman, 2014) and applied specifically to the male/female distribution in the military.

[e] Based on estimates of prevalence from a Massachusetts study (Gates, 2011).

NOTE: Hormone therapy is person-level; surgery statistics are counts of surgeries, and one person may have multiple surgeries.

ing the information on which these estimates rely. As stated previously, these concerns include both a reliance on prevalence estimates from just two states and a reliance on data from the NTDS, which were not collected from a random sample. Our utilization estimates were taken primarily from three sources:

- private health insurance utilization data on annual rates of enrollee transgender-related health care utilization in health insurance plans that cover transition-related health care, as reported by Herman (2013b)
- private health clinic data showing estimates of the rates of penectomies and bilateral mastectomies in the U.S. population in 2001, as reported by Horton (2008)[5]

---

[5] A penectomy is the surgical removal of the penis. A bilateral mastectomy is the surgical removal of both breasts.

- Veterans Health Administration (VHA) claims data, which were used to calculate prevalence and incidence rates of gender identity disorder (now referred to as GD in DSM-5) from 2006 to 2013, as reported by Kauth et al. (2014).

Each of these data sources provides information on a different outcome, which makes understanding the results more complicated. However, collectively, the information taken from these three studies provides a broad, useful picture regarding potential gender transition–related health care utilization in the AC population. In the following sections, we review each of these studies in detail, identify key estimates from each, and apply the estimates to the AC population identified in Table 3.2 in Chapter Three.

Private Health Insurance Utilization Estimates

Herman (2013b) reports on the experiences of 34 employers that provided gender transition–related health care benefits to their employees and dependents via their health insurance plans. This study specifically reports on the annual number of enrollees who accessed "transition-related care." This information is derived from health insurance claims data and thus is dependent on the treatments that were covered by the health insurance companies.[6] The firms surveyed typically covered major gender transition–related surgeries and hormone therapy, but they varied in their coverage of other transition-related treatments, such as vocal cord surgery.[7]

Firms reviewed by Herman (2013b) also typically did not report information on the number of dependents covered but included dependents in their utilization estimates. Data from several sources (e.g., Sonier et al., 2013; Gould, 2012) imply an approximate average one-to-one ratio of employees to dependents in privately insured firms in the United States. Thus, not accounting for the role of dependents in these utilization estimates would overstate utilization by approximately 100 percent.[8] For

---

[6] If firms do not cover particular treatments, it is not possible to file a claim for reimbursement. If individuals in these firms utilized services that were not covered, thus paying for treatments out of pocket or through some other form of health insurance, these utilization estimates will be biased downward.

[7] One hundred percent of firms covered major gender transition–related surgeries, including hysterectomy, oophorectomy, metoidioplasty, phalloplasty, urethroplasty, vaginectomy, orchiectomy, vaginoplasty, labiaplasty, and clitoroplasty. Ninety-two percent of firms covered bilateral mastectomy for female-to-male patients, but only 59 percent covered female-to-male chest reconstruction, and only 59 percent covered male-to-female augmentation mammoplasty (breast augmentation). All firms covered hormone therapies, specifically estrogen, progesterone, spironolactone, and testosterone.

[8] We used two different data sources to determine the typical number of dependents covered by the main policyholder in private health insurance firms in the United States. First, we used information from the Robert Wood Johnson Foundation on the number of people who are covered by employer-sponsored health insurance and are the main policyholders and on the number of people who are covered by employer-sponsored health insurance and are dependents. Using these figures, we estimated a 1-to-0.99 policyholder-to-dependent ratio in employer-sponsored private health insurance. The Economic Policy Institute also reports information on this question using data from the U.S. census Current Population Survey. Using this information, we calculated a policyholder-to-dependent ratio of 1 to 0.94.

firms that did not provide information on dependents, we imputed a one-to-one ratio of employees to dependents to identify the total number of enrolled individuals in a given health plan.

Table 4.4 presents the information from Herman (2013b) on the utilization of gender transition–related care in private health insurance firms. The first column shows available information on the identity of the firm. The second describes the number of firms in each category for which we had utilization estimates. The third contains our estimates regarding the total number of enrollees and dependents from all firms in that category. For confidentiality reasons, some surveyed data sources report only ranges for the number of employees in a firm. Therefore, we used the midpoint of the range to impute the number of employees in a particular firm, then assigned the total number of dependents based on this employee value. For example, we had utilization data from two firms in the "private 1,000–9,999 employees" category. Since we assume the midpoint value for firm size, this implies that there are 5,000 employees in each firm, or 10,000 total employees across the two firms. Assuming a one-to-one employee-to-dependent ratio implies an additional 10,000 covered individuals, resulting in a combined total of 20,000 enrollees.

The estimates presented in Table 4.4 indicate that utilization rates range from an annual low of zero individuals per 1,000 enrollees to an annual high of 0.064 individuals per 1,000 enrollees. To obtain a combined estimate of the different values, we constructed a weighted average using the existing utilization estimates, weighting by the number of covered individuals that generated each of the estimates in Table 4.4. A weighted average of all the estimates results in an overall utilization estimate of 0.0396 individuals per 1,000 enrollees.

Table 4.4
Enrollee Utilization of Gender Transition–Related Benefits in Private Health Insurance Firms

Private and Public Firms	Number of Firms	Total Contribution (enrollees + dependents)	Individual Claimants per 1,000 Enrollees
Private, fewer than 1,000 employees	1	1,000	0.0000
Private, 1,000–9,999 employees	2	20,000	0.0540
Private, 10,000–49,000 employees	5	250,000	0.0220
City and County of San Francisco	NA	80,000	0.0640
University of California	NA	100,000	0.0620
Weighted average per 1,000 enrollees			0.0396

SOURCE: Data from Herman, 2013b.

We conducted two sets of calculations using these estimates. First, we used the lowest non-zero utilization figure (0.022 claimants per 1,000 enrollees);[9] then, we used the weighted average calculation of 0.0396 per 1,000 enrollees. Applying the 0.022 claimants per 1,000 figure to the AC population of 1,326,273 implies that 29 AC service members would receive gender transition–related care annually. Applying the weighted average estimate of 0.0396 per 1,000 enrollees to the AC population implies that 53 service members would receive gender transition–related care annually.

Sensitivity Analyses

We also conducted two additional sensitivity analyses to determine the full potential scope of gender transition–related health care utilization in the AC. A key consideration when applying estimates from civilian populations to the military is that the underlying male/female distribution in the AC is different, with 85 percent of the AC population being male (versus approximately 50 percent in the civilian population). Studies suggest that the prevalence of transgender individuals is higher in the male population than in the female population (APA, 2013; Horton, 2008; Gates, 2011; Grant et al., 2011), so applying civilian estimates directly to the AC would underestimate the true utilization rates.

Accurately accounting for this issue required sex-specific utilization estimates that we could then multiply with the male/female AC distribution (85 percent male, 15 percent female). Unfortunately, we could not identify any sex-specific utilization estimates in the available private health insurance data; the aggregate cost and utilization estimates that we were able to identify already included underlying prevalence differences between the sexes. We posited that utilization would be twice as large for male-to-female transitions than for female-to-male transitions based on an assumption of linearity between transgender prevalence, for which we have sex-specific estimates, and total utilization (Horton, 2008).

Combining this assumption about differing utilization rates with the fact that the male/female labor force participation in the civilian population is close to 50 percent male and 50 percent female, we were able to solve for the sex-specific utilization estimates implied by the aggregate lower-bound (0.022) and weighted average (0.0396) values. Solving for the sex-specific utilization estimates in this manner, for the 0.022 aggregate estimate, we estimated a utilization rate of 0.0293 per 1,000 natal male enrollees and a utilization rate of 0.0146 per 1,000 natal female enrollees.[10] Similarly, for the 0.0396 weighted average figure, solving for the natal sex–specific utiliza-

---

[9] The unadjusted version of this figure (0.0044 percent) was also used in Belkin (2015) to estimate health care utilization in the military.

[10] The equation we solved to calculate the natal male–specific and natal female–specific utilization rates is as follows: $0.5(x) + 0.5(2x) = 0.022$. In this equation, the variable x is the natal female–specific utilization rate, and solving for x results in a value of 0.0146. Since the natal male–specific utilization rate is assumed to be twice the natal female rate, it equals 0.0293.

tion estimates, we identified a utilization rate of 0.0528 per 1,000 natal male enrollees and a utilization rate of 0.0264 per 1,000 natal female enrollees.

Applying these solved sex-specific estimates to the AC male/female distribution (1,125,581, or 85 percent male, versus 200,692, or 15 percent female) increased our initial lower-bound estimate of claimants from 29 to 36 and increased our estimate from applying the weighted average from 53 to 65.

Finally, the sociology and psychology literature speculates that there is a higher transgender prevalence in the military compared with the civilian population (G. Brown, 1988). Gates and Herman (2014) also calculated that transgender prevalence in the military is approximately twice the civilian prevalence (Gates, 2011; Gates and Herman, 2014).[11] Although we believe that the current body of empirical evidence validating this theory is weak, we take it seriously and consider the possible implications for transition-related health care utilization in the military. Assuming that transgender prevalence in the military is twice the transgender prevalence in the civilian population, and, again, assuming a direct relationship between prevalence and utilization, this would inflate our male/female distribution-adjusted estimates of individuals receiving transition-related care annually from 36 to 72, and from 65 to 129 in the AC. Table 4.5, which summarizes the results from applying the private health insurance estimates to the AC population, allows for a comparison of the different estimates.

Private Health Clinic Estimates

A second source of information regarding gender transition–related health care utilization comes from a survey of surgical clinics conducted by Horton (2008). In 2001, Horton surveyed all major clinics in the United States known to provide transition-related care to determine the number of penectomies and bilateral mastectomies performed on transgender patients. Table 4.6 reports surgery incidence estimates broken out by male-to-female transitions and female-to-male transitions. The third column shows estimates using clinic-reported data only. Horton also developed lower- and upper-bound estimates via assumptions regarding treatment counts for clinics with missing data, and these numbers are reported in the second and fourth columns of Table 4.6.[12] These data were collected in 2001 and coverage of gender transition-related benefits have increased over time, so it is also reasonable to assume that surgical tran-

---

[11] As stated previously, Gates and Herman (2014) used estimates from the NTDS and Gates (2011) for a transgender prevalence of 0.3 percent. That study also used data on history of military service in the U.S. population from the American Community Survey to estimate transgender prevalence in the military. Data from the National College of Health Administration show that military experience was significantly higher among transgender individuals than among those who did not identify as transgender (9.4 percent versus 2.1 percent; Blosnich, Gordon and Fine, 2015). However, data were collected from only 51 institutions, and the response rate for the survey was only 20 percent, which again raises questions regarding the validity of the estimates.

[12] Horton generated upper- and lower-bound estimates by assigning the largest and smallest surgical counts in the data to the clinics with missing values.

Table 4.5
Utilization Estimates from Applying Private Health Insurance Parameters

Annual Individual Claimants	Estimate from the Literature	Estimates Using Private Employer Data		
		Baseline	Sensitivity Analysis 1[a]	Sensitivity Analysis 2[b]
Active component, lower-bound estimate	0.022 claimants per 1,000 individuals	29	36	72
Active component, weighted average estimate	0.0396 claimants per 1,000 individuals	53	65	129

NOTES: Each cell in the "Estimates Using Private Employer Data" columns represents a unique prediction for utilization in the AC population. In the second column of the table, we describe the estimate from the literature that is applied to the AC population. See the text for details on each of the calculations.

[a] Sensitivity Analysis 1: We calculated a set of estimates that accounted for differences in the male/female distribution between the civilian and AC populations.

[b] Sensitivity Analysis 2: We calculated a set of estimates that accounted for differences in the male/female distribution between the civilian and AC populations and the possibility that transgender prevalence is twice as high in the military population as in the civilian population.

Table 4.6
Incidence of Penectomies and Bilateral Mastectomies Performed on Transgender Individuals

Transition Type	Incidence Estimates (%)		
	Low	Clinic-Reported Data	High
Male-to-female	0.00048	0.00053	0.00103
Female-to-male	0.00020	0.00030	0.00084

SOURCE: 2001 data from Horton, 2008.
NOTE: The table includes data on penectomies and bilateral mastectomies only.

sitions have also increased over time. Thus, these utilization rates of penectomies and bilateral mastectomies should be considered lower-bound estimates.

Applying these estimates to the AC male/female distribution results in low, medium, and high annual estimates of 5.8, 6.6, and 13.2 AC service members receiving these two surgeries, respectively. We reiterate here that these estimates are not directly comparable to the private health insurance estimates presented in the previous section because these estimates apply to only two specific procedures, while the private health insurance estimates include any gender transition–related procedures that private health insurance firms cover. One would expect estimates for two specific surgeries from 2001 to be lower than estimates generated from the private health insurance system in the later 2000s. Indeed, they are, but it is more difficult to make other direct comparisons between these two estimates, given the private health insurance utilization data presented in Herman (2013b).

Veterans Health Administration Estimates

In this analysis, we used VHA data to calculate the expected annual incidence of gender identity disorder (the condition now known as GD in the DSM-5) in the AC population. As described previously, those with a gender identity disorder diagnosis are a subset of transgender individuals. Kauth et al. (2014) used VHA health claims data to identify incidence rates of new diagnoses. They also calculated prevalence rates of gender identity disorder in each year using previous yearly incidence rates. Because 2006 was the first year in their data set, the prevalence rate in the first year of their data is equivalent to the incidence rate. In the years after 2006, the prevalence rate is essentially a running total of the incidence rates in the previous years added to the most recent incidence rates.

The data in Table 4.7 imply that the incidence of gender identity disorder increased from 3.5 of 100,000 enrollees in FY 2006 to 6.7 of 100,000 enrollees in FY 2013 among veterans who use VHA health care (Kauth et al., 2014). Before applying these estimates to the AC population, we note two important points with respect to the analyses in Kauth et al. (2014). First, because the prevalence rate is simply a running total of new cases diagnosed since the first year of the study's data (2006), adding years of data prior to 2006 would mechanically increase the prevalence estimates. Thus, Kauth et al.'s prevalence calculations are a lower-bound for the total gender

Table 4.7
Prevalence and Incidence of Gender Identity Disorder Diagnoses in VHA Claims Data

Fiscal Year	New Diagnosis Rate (%)	Prevalence (%)
2006	0.0035	0.0035
2007	0.0034	0.0068
2008	0.0034	0.0098
2009	0.0038	0.0131
2010	0.0046	0.0172
2011	0.0051	0.0217
2012	0.0060	0.0270
2013	0.0067	0.0329

SOURCE: Kauth et al., 2014.

NOTE: The authors calculated new cases diagnosed and total existing cases in a given year based on the entirety of the data since 2006.

identity disorder prevalence rate in this population. Second, estimates based on claims data will likely be lower-bound estimates of incidence and prevalence, since individuals are identified only if they interact with the health care system for reasons related to gender identity disorder. These two caveats should be kept in mind when interpreting the extrapolations here.

Applying estimates from the 2013 data in Table 4.7 to the AC population, one would expect approximately 90 new cases of gender identity disorder each year and that approximately 440 AC service members would be diagnosed with this condition. Although the male/female distribution in the VHA system mirrors that of the AC, veterans who use VHA health care services may have lower socioeconomic and health status than veterans who do not use VHA health care, other military retirees, and AC service members. The VHA population also differs by age and, potentially, by other unmeasured characteristics related to underlying health status. For these varied reasons, these estimates may not be generalizable to the military population overall.

Summarizing the Estimates

Table 4.8 summarizes the key results after applying the estimates from the various data sets to the AC and SR populations. The largest estimate—270 treatments (surgeries and hormone therapies)—was calculated by combining the upper-bound population-level transgender prevalence estimate from Massachusetts with information from the NTDS data on the age of those receiving common transition-related treatments. When applied to the AC population, estimates from VHA and the private health insurance literature imply that only 30–90 AC service members will receive some type of gender transition–related treatment annually.

To understand the full implications of our estimates regarding the expected annual number of AC service members likely to obtain gender transition–related care, in Figure 4.1 we compare the above utilization estimates with the number of AC service members who self-reported visiting a mental health care provider in a given year (21 percent) and the number of AC service members who visited a mental health care specialist in a given year (7 percent; Hoge et al., 2006; McKibben et al., 2013). We chose this outcome because mental health care among military populations is an important, well-studied topic, and data were readily accessible for us to conduct the comparison. The mental health care utilization estimates represent unique service members accessing health care; thus, they compare most directly to the estimates using the private health insurance data and the NTDS hormone therapy estimates. For clarity's sake, we do not present all of the private health insurance and NTDS hormone therapy estimates in Figure 4.1. We do include the smallest, middle, and largest estimates using the private health insurance data and the largest hormone therapy estimate drawn from the NTDS data.

Table 4.8
Annual Gender Transition–Related Treatment Estimates from All Data Sources

Estimate Type	Active Component			Selected Reserve		
	Hormone Treatment	Surgical Treatments	All Treatments	Hormone Treatment	Surgical Treatments	All Treatments
Prevalence-based estimates (using NTDS data)						
Annual treatments based on CA study estimate (0.1%)	30	25	55	15	15	30
Annual treatments based on combined, population-weighted, gender-adjusted rate (0.19%)	50	45	95	25	30	55
Annual treatments based on MA study estimate (0.5%)	140	130	270	70	80	150
Utilization-based estimates						
Private health insurance annual individual claimants (0.022 per 1,000)	NA	NA	29	NA	NA	20
Private health insurance annual individual claimants (0.0396 per 1,000)	NA	NA	53	NA	NA	30
VHA-based annual new diagnoses (0.0067%)	90	NA	NA	60	NA	NA
Clinical utilization of penectomies and bilateral chest surgeries (0.0005%)	NA	10	NA	NA	5	NA

SOURCE: RAND analysis.

As Figure 4.1 shows, our estimates of the number of AC personnel who will use the gender transition–related health care benefits are overwhelmingly small compared with the number of AC personnel who access mental health treatment. Overall, based on our calculations, we expect annual gender transition–related health care to be an extremely small part of overall health care provided to the AC population.

Figure 4.1
Comparison of Annual Estimated Gender Transition–Related Health Care Utilization and Mental Health Care Utilization, Active Component

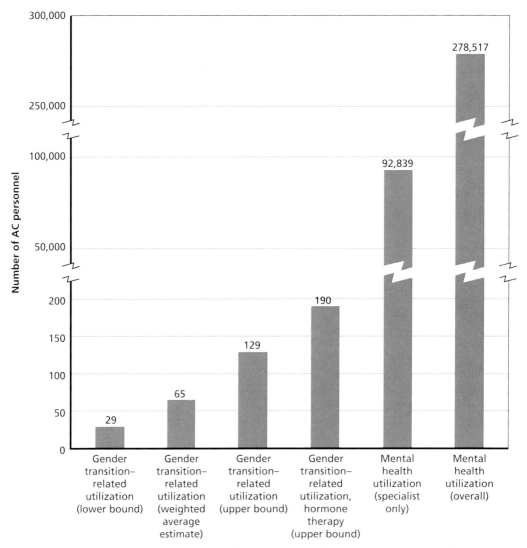

SOURCE: RAND analysis. Utilization rates in the figure are derived from both the prevalence-based and utilization-based approaches presented in Table 4.8.

NOTES: The non–hormone therapy transgender utilization estimates are from the application of estimates from the private health insurance data. The hormone therapy upper-bound transgender utilization estimate is from calculations using the NTDS data.

RAND RR1350-4.1

CHAPTER FIVE

What Are the Costs Associated with Extending Health Care Coverage for Gender Transition–Related Treatments?

In this chapter, we provide estimates for the costs associated with extending health care coverage for gender transition–related treatments. We focused on transgender service members in the AC because they have uniform MHS access. We did not include reserve-component service members in our analyses, but their MHS utilization and the associated cost will be negligible, given their highly limited military health care eligibility. Likewise, we did not include retirees or dependents in the cost analyses because we did not have information on age and sex distribution within these beneficiary categories. Some of these beneficiary categories also have limited eligibility for health care provided through MTFs and may receive their health care through TRICARE coverage in the purchased care setting or through other health insurance plans. Given these unknowns, it was only feasible to estimate the costs of gender transition–related care for AC service members; however, we recommend expanding these analyses in the future to include reserve-component members, as well as all individuals eligible for treatment under TRICARE. For the following analyses, we used demographic characteristics of the 2014 AC population to estimate the cost of providing such services.

Private Health Insurance Cost Estimates

To determine the potential costs of covering gender transition–related health care for transgender service members, we collected information on private health insurers' experiences with covering this care from two sources (Herman, 2013b; State of California, 2012). These actuarial estimates represent the expected increase in health care costs from covering a new set of treatments or a new group of beneficiaries. If employers decide to provide coverage for a particular treatment, these actuarial estimates are translated into premium increases for covered employees. These estimates should be thought of as the expected costs of extending coverage for gender transition–related care to transgender AC service members. Moreover, we note that the military may already be incurring the cost of some transgender treatments, as some patients and their providers use "omissions and ambiguities" to acquire needed care (Roller, Sedlak, and Draucker, 2015, p. 420). For example, a currently serving female-to-male patient who had undergone a hysterectomy reported taking only the testosterone and not the estrogen prescribed as part of hormone therapy with his endocrinologist's knowledge and tacit support, while another was trying to get breast reduction surgery due to back pain rather than GD (Parco, Levy, and Spears, 2015, pp. 235–236).

Table 5.1 presents available data from public employers and private firms on the actuarial costs of covering gender transition–related care. It identifies the particular institution, the number of employees and dependents covered, and the identified premium increases due to expanding benefits.

Data from Table 5.1 show, generally, that the actuarial estimates of providing benefits for gender transition–related care increased total premiums (employee + employer share) by only a small fraction of a percent—and, in the most extreme cases, by only approximately 1 percent. Taking a weighted average of most of the information,[1] we estimated that extending insurance coverage to transgender individuals would increase health care spending by 0.038 percent. Applying this figure to total AC health care spending of $6.27 billion,[2] we find that covering gender transition–related care will increase AC health care spending by approximately $2.4 million (see Table 5.2).

The data in Table 5.1 suggest that the University of California, with 100,000 enrollees in its health plan, is one of the key drivers of the 0.038-percent weighted

Table 5.1
Actuarial Estimated Costs of Gender Transition–Related Health Care Coverage from the Literature

Public Employer Data	Actuarially Calculated Premium Increase	Total Contribution (employees + dependents)
City of Seattle	0.19% increase in health care budget	23,090
City of Portland	0.08% increase in health care budget	18,000
City of San Francisco	0% increase in health care budget	100,000
University of California	0% increase in health care budget	100,000
Private Employer Data	**Estimate**	**Total Contribution (employees + dependents)**
22 firms	Many employers reported no actuarial costs to adding benefit; estimates range from 0 to 0.2%	Mix of firm sizes
2 firms	Approximately 1% increase in premiums	5,800
1 firm	Much less than 1% increase in premium	77,000

SOURCE: Estimates are from Herman, 2013b, and State of California, 2012.

---

[1] We did not use information about the firm with 77,000 enrollees because it is not clear what "much less than 1 percent" implies with respect to the premium increase.

[2] Pharmaceutical and direct and purchased care inpatient and outpatient data calculated from TRICARE costs in Defense Health Agency, 2015.

average result. In addition to the actuarial increases, the University of California also reported a realized increase in health care spending of 0.05 percent, so we recalculated the weighted average figure by replacing the 0-percent estimate with the 0.05 percent estimate. This new calculation raised the overall cost estimate from 0.038 percent to 0.054 percent, or from $2.4 million to $3.4 million when applied to the AC. To summarize, our baseline estimates regarding expected gender transition–related health care costs in the AC are between $2.4 million and $3.4 million.

Sensitivity Analyses

To understand the potential full range of cost effects in the AC population, we conducted two additional sensitivity analyses similar to those described for our utilization ranges in Chapter Four. We used these sensitivity analyses to account for the skewed male/female distribution in the military population and for the possibility that transgender prevalence is higher in the military population. As in the utilization case, we were not able to identify any sex-specific effects on the premium increases. Thus, as in our utilization analysis, we assume that cost estimates are linearly related to prevalence,[3] and cost estimates for male-to-female transitions are twice the cost estimates for female-to-male transitions. Using this relationship, we again calculated natal male– and natal female–specific estimates from the aggregate estimates.

Given the assumption about differing cost effects, we calculated a natal male–specific cost estimate of 0.05 percent and a natal female–specific cost estimate of 0.025 percent for the aggregate premium estimate of 0.038 percent. Applying these sex-specific estimates to the AC male/female distribution increased our initial premium estimate from 0.038 percent to 0.047 percent. A similar calculation can be performed for our realized cost estimate of 0.054 percent. Assuming that gender transition–related health care costs are twice as large for male-to-female transitions as for female-to-male transitions, we calculated a natal male–specific cost effect of 0.072 percent and a natal female–specific cost effect of 0.036 percent. Applying these sex-specific estimates to the AC male/female distribution increased our initial premium estimate from 0.054 percent to 0.067 percent. Applying these newly calculated health care costs to the 2014 AC health care expenditures ($6.27 billion) increased our estimate of costs from the initial range of $2.4–3.4 million to a range of $2.9–4.2 million.

Finally, as noted previously, Gates (2011) and Gates and Herman (2014) calculated that transgender prevalence in the military is approximately twice that in civilian

---

[3]   We also note that built into this linearity assumption and how it is applied in the two sensitivity analyses is the assumption that the cost of male-to-female transitions is the same as the cost of female-to-male transitions. Since there is no sex-specific information in the private health insurance cost data, the validity of the cost per case being equivalent is unknown. Padula, Heru, and Campbell (2015) estimated that a male-to-female surgical case is 33 percent more expensive than a female-to-male surgical case, but these estimates were not based on private employer data, so we did not directly incorporate this result into our calculations.

populations. Assuming that this estimate is valid, and, again, assuming that health care costs are linearly related to underlying prevalence, this would increase the above calculated value of $2.9 million to $5.8 million and the calculated value of $4.2 million to $8.4 million. Table 5.2 summarizes the results from the calculations described in this section.

To better understand the relative importance of our estimates regarding expected AC annual gender transition–related health care spending, we compared our cost estimates to the MHS spending on mental health in 2012 and to total AC health care spending in FY 2014. As Figure 5.1 shows, gender transition–related health care spending is expected to be extremely small compared with MHS spending on mental health (Blakely and Jansen, 2013) and overall AC health care expenditures (Defense Health Agency, 2015).

Summarizing the Estimates

A direct application of estimates from the private health insurance system implies a baseline spending range between $2.4 million and $3.4 million for AC gender transition–related health care. Sensitivity analyses that attempt to account for the fact that the male/female distribution in the AC population skews more heavily male than the civilian population and that transgender prevalence might be higher in the military increase this initial range to $5.8 million to $8.4 million. The implication is that even in the most extreme scenario that we were able to identify using the private health insurance data, we expect only a 0.13-percent ($8.4 million out of $6.2 billion) increase in AC health care spending.[4]

Table 5.2
Estimated Annual MHS Costs of Gender Transition–Related Health Care, Active Component

Analysis Type	Calculations Using Only Actuarial Premium Estimates 0.038% (actuarial)	Calculations Using Actuarial Premiums and Realized Values 0.054% (actuarial + realized)
Baseline	$2.4 million	$3.4 million
Sensitivity analysis 1: Adjusts for the male/female distribution in the AC population	$2.9 million	$4.2 million
Sensitivity analysis 2: Adjusts for the male/female distribution in the AC population and the assumption that transgender prevalence is twice as high in the military compared to the civilian population	$5.8 million	$8.4 million

SOURCE: RAND analysis.

---

[4]   AC beneficiaries make up less than 15 percent of total TRICARE beneficiaries (Defense Health Agency, 2015).

Figure 5.1
Gender Transition–Related Health Care Cost Estimates Compared with Total Health Spending, Active Component

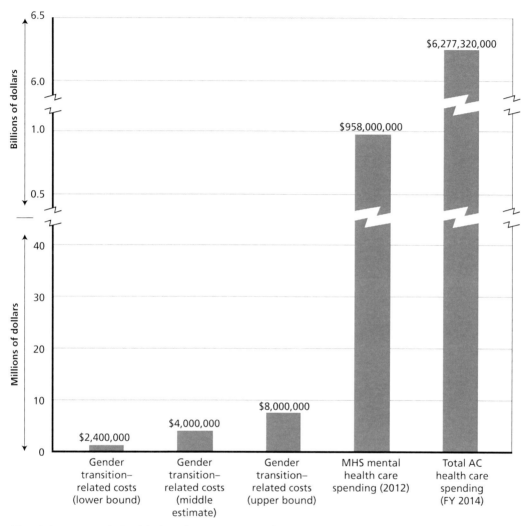

SOURCES: RAND analysis; Blakely and Jansen, 2013; Defense Health Agency, 2015. Estimates of premium increased and realized costs are reported in Table 5.1.

NOTES: The lower-bound estimate refers to premium increases only. The middle estimate includes premium increases and realized costs after adjusting for male/female distribution in the military. The upper-bound estimate includes premium increases and realized costs after adjusting for male/female distribution in the military and assuming the prevalence rate of transgender individuals in the military is twice that of civilian populations.

RAND *RR1350-5.1*

CHAPTER SIX

What Are the Potential Readiness Implications of Allowing Transgender Service Members to Serve Openly?

As DoD considers whether to allow transgender personnel to serve openly and to receive transition-related treatment during the course of their military service, it must consider the implications of such a policy change on the service members' ability to deploy and potential reductions in unit cohesion. In prior legal challenges to the transgender military discharge policy, DoD has expressed concern that the medical needs of these service members would affect military readiness and deployability. To address these concerns, this chapter provides estimates of the potential effects on force readiness from a policy change allowing these service members to serve openly.

A critical limitation of such an assessment is that much of the current research on transgender prevalence and medical treatment rates relies on self-reported, nonrepresentative samples. Thus, the information cited here must be interpreted with caution because it may have varying degrees of reliability. In addition, to estimate effects on readiness, we focused on transgender personnel in the AC and SR only. We did not include the Individual Ready Reserve because of the lack of publicly available, detailed demographic information. We used the same approach that applied to our analysis of health care utilization, applying both the prevalence-based and utilization-based approaches to force size. We note that the prevalence-based approach was the only approach that allowed us to estimate the number of transgender service members who may seek to live and work as their target gender. Transition does not necessarily imply the use of medical treatments, and we emphasize that some of these service members may still require accommodations in terms of housing and administrative functions (e.g., military identification cards, restrooms).

Impact on Ability to Deploy

The most salient and complex issue in allowing transgender personnel to serve openly is how DoD should regulate and manage operational deployment requirements for these personnel in the context of their transition to their target gender.

Pre-Transition

If transgender personnel are allowed to serve openly prior to transition, DoD will need to establish policies on when individuals may use the uniforms, physical standards, and facilities (e.g., barracks, restrooms) of their target gender. Additionally, DoD will need to clarify policies related to qualifications for deployment. Current deployment rules suggest that to qualify for deployment, individuals with diagnosed mental health disorders must show a "pattern of stability without significant symptoms or impairment for at least three months prior to deployment."[1] Ensuring appropriate screening will be critical to minimizing any mental health–related readiness issues. Secondary prevention measures prior to deployment, such as screening for GD, may be needed to ensure a pattern of stability and readiness for deployment.

During Transition

DoD would also need to determine when transitioning service members would be able to change uniforms and adhere to the physical standards of their target gender, as well as which facilities and identification cards they will use. Other countries have found that, in some cases, it may be necessary to restrict deployment of transitioning individuals to austere environments where their health care needs cannot be met. Deployment restrictions may also be required for individuals seeking medical treatment, including those seeking hormone therapy and surgical treatments.

We detail the constraints associated with transition-related medical treatments in Table 6.1. These constraints typically include a postoperative recovery period that would prevent any work and a period of restricted physical activity that would prevent deployment. The rightmost column of Table 6.1 presents the estimated number of non-deployable days we used to estimate the readiness impact. We note that these estimates do not account for any additional time required to determine medical fitness to deploy. Army guidelines, for example, do not permit deployment within six weeks of surgery. Nevertheless, there may be a significant difference between the estimated availability to deploy and the actual impact on deployability, as it is possible that transgender service members would time their medical treatments to minimize the effect on their eligibility to deploy.[2]

In addition to an expected, short-term inability to deploy during standard postoperative recovery time, some individuals experience postoperative complications that would render them unfit for duty. For instance, among those receiving vagino-

---

[1]   Detailed guidance is provided in a memorandum from the Office of the Assistant Secretary of Defense for Health Affairs, 2013, p. 2.

[2]   See for example, Personnel Policy Guidance Tab A (known as PPG-TAB A) that accompanies the medical guidelines document MOD TWELVE, Section 15.C, which articulates the minimal standards of fitness for deployment to the U.S. Central Command area of responsibility (U.S. Central Command, 2013).

plasty surgery, 6–20 percent have complications.[3] This implies that between three and 11 service members per year would experience a long-term disability from gender reassignment surgery. Among those receiving phalloplasty surgery, as many as 25 percent experience some medical complications (Elders et al., 2014).

Table 6.1
Gender Transition–Related Readiness Constraints

Transition Type and Treatment	Recovery Time	Leave and Deployment Implications	Estimated Nondeployable Days
Male-to-Female			
Hormone therapy only	Long-term, no recovery required	None (pending accommodations)	N/A
Augmentation mammoplasty	1 week no work, 4–6 weeks restricted physical activity	Up to 14 days medical leave, up to 60 days medical disability	75
Genital surgery (orchiectomy, vaginoplasty)	4–6 weeks no work, 8+ weeks restricted physical activity	Up to 45 days medical leave, up to 90 days medical disability	135
Female-to-Male			
Hormone therapy only	Long-term, no recovery required	None (pending accommodations)	N/A
Chest surgery	1 week no work, 4–6 weeks restricted physical activity	Up to 14 days medical leave, up to 60 days medical disability	75
Hysterectomy	2 weeks no work, 4–8 weeks restricted physical activity	Up to 21 days medical leave, up to 90 days medical disability	111
Genital surgery (metoidioplasty, phalloplasty)	2–4 weeks no work, 4–6 weeks restricted physical activity	Up to 21 days medical leave, up to 60 days medical disability	81

SOURCES: Treatment times based on RAND research compiled for this study. Estimates of numbers of treatments based on rates in Gates, 2011. Estimated nondeployable days based on RAND calculations using FY 2014 data from DoD, 2014.

NOTES: The total population in the table includes AC and SR personnel. Estimates of treatments are non-unique per person. Individuals may (and likely will) seek multiple treatments simultaneously. As such, deployment days are measured per treatment, not per individual. Estimates of nondeployable days do not include estimated delays generated by Medical Evaluation Board/Physical Evaluation Board review, which may be required depending on service rules.

---

[3] According to Elders et al. (2014, p. 15), summarizing findings from 15 studies, "2.1 percent of patients had rectal-vaginal fistula, 6.2 percent with vaginal stenosis, 5.3 percent had urethral stenosis, 1.9 percent with clitoral necrosis, and 2.7 percent with vaginal prolapse," and approximately 2.3 percent of patients experienced complications after vaginoplasty.

Taking the estimates for treatment and recovery time, we then applied the standards for leave and restricted physical activity.[4] We applied the recovery times and translated those into nondeployable days separated into medical leave, in which the service member is off the job, and medical disability, in which the service member can be at work but is subject to restricted physical requirements (e.g., no physical training, no heavy lifting). This provided us with the total number of nondeployable days per treatment type. We scaled this estimate by the number of days an individual can be deployed per year. For the AC, we assumed this to be 330 days per year (allowing 30 days of leave plus five days of processing time).[5] For the SR, we assumed 270 days per year (which allows nine months of deployment time). We counted each treatment separately and applied the number of treatments by treatment type shown in Table 6.1.

Note that because individuals may seek multiple treatments, sometimes at the same time, this number is not the same as the total number of individuals who will be nondeployable. Therefore, the estimates presented in Table 6.2 should be considered an upper bound in each category. Moreover, the prevalence-based estimates are significantly larger than the utilization-based estimates as shown in Table 4.8. Using the prevalence-based approach, we found that between eight and 43 of the available 1.2 million labor-years in the AC may be unavailable for deployment.[6] The combined, population-weighted, and gender-adjusted estimate implies that about 16 labor-years from the AC and about 11 labor-years from the SR may be nondeployable. This represents 0.0015 percent of available deployable labor-years across the AC and SR.

These estimates are based on surgical take-up rates ranging from 25 to 130 per year in the AC, with 55–270 total treatments, including hormone treatments. Similarly, the prevalence-based estimates imply 15–80 surgical treatments per year in the SR, with between 30 and 150 total treatments, including hormone therapy.

The utilization-based approach implies many fewer treatments. Although we could not estimate the impact on labor-years because we did not have information on specific treatments, based on usage rates in California, the utilization-based approach implies 30–50 total treatments, including surgeries and hormone therapy. Evidence from the VHA suggests that 90 service members in the AC and 50 in SR are diagnosed with GD in any given year. Such a diagnosis would be a prerequisite for any surgical treatments, suggesting that true utilization rates in the military may be significantly lower than suggested by the prevalence-based approach.

We caution that our labor-year estimates also likely overcount actual nondeployable time because our estimate captures "availability to deploy," rather than the deploy-

---

[4] For reference, we used the Army Regulation 40-501 (revised 2011), which governs leave and disability, and the Navy Medical Policy 07-009 (2007), which provides guidance on pre-clearance, accommodations for deployment readiness, and additional requirements in the U.S. Central Command area of operations.

[5] We based this estimate on Army Regulation 600-8-101 (2015).

[6] We define a labor-year as the amount of work done by an individual in a year.

Table 6.2
Estimated Number of Nondeployable Man-Years Due to Gender Transition–Related Treatments

Component	Total Labor-Years Available (FY 2014)	0.1%[a] (CA study)	0.16%[b] (combined, population-weighted CA + MA studies)	0.19%[c] (gender-adjusted rate)	0.37%[d] (twice gender-adjusted rate)	0.5%[e] (MA study)
			Estimated Number of Nondeployable Labor-Years			
Active	1,199,096	8.2	13.7	16.2	32.3	42.8
Selected Reserve	615,446	5.9	9.9	10.7	21.3	29.9

SOURCES: Estimates for nondeployable labor-years are based on RAND calculations using FY 2014 data from DoD, 2014.

[a] Based on estimates of prevalence from a California study (Conron, 2012).

[b] Based on weighted average of studies from California and Massachusetts, weighted by relative population sizes in each state.

[c] Based on weighted average of studies from California and Massachusetts, weighted by relative population sizes in each state and applied specifically to the male/female distribution in the military components.

[d] Based on estimates of prevalence from NTDS, Gates (2011), and the American Community Survey (Gates and Herman, 2014) and applied specifically to the male/female distribution in the military.

[e] Based on estimates of prevalence from a Massachusetts study (Gates, 2011).

ment impact itself. This difference comes from three key assumptions that we make to calculate these estimates: (1) service members who are seeking treatment will also be deployed; (2) service members who are seeking treatment cannot time those treatments to avoid affecting their deployment eligibility; and (3) service members seek only one treatment at a time rather than having multiple treatments at the same time, which would allow concurrent (rather than sequential) recovery times. Thus, it is likely that a service member's care would have a substantial overall impact on readiness only if that service member worked in an especially unique military occupation, if that occupation was in demand at the time of transition, and if the service member needed to be available for frequent, unpredicted mobilizations.

Post-Transition

Having completed medical transition, a service member could resume activity in an operational unit if otherwise qualified. As in other cases in which a service member receives a significant medical treatment, DoD should review and ensure that any longer-term medical care or other accommodations relevant to the transgender service member's specific medical needs are addressed.

Impact on Unit Cohesion

A key concern in allowing transgender personnel to serve openly is how this may affect unit cohesion—a critical input for unit readiness. The underlying assumption is that if service members discover that a member of their unit is transgender, this could inhibit bonding within the unit, which, in turn, would reduce operational readiness. Similar concerns were raised in debates over whether to allow gay and lesbian personnel to serve openly (Rostker et al., 1993; RAND National Defense Research Institute, 2010), as well as whether to allow women to serve in ground combat positions (Schaefer et al., 2015; Szayna et al., 2015). Evidence from foreign militaries and surveys of the attitudes of service members have indicated that this was not the case for women or for lesbian and gay personnel (Schaefer et al., 2015; Harrell et al., 2007; RAND National Defense Research Institute, 2010). In examining the experiences of foreign militaries, the limited publicly available data we found indicated that there has been no significant effect of openly serving transgender service members on cohesion, operational effectiveness, or readiness. (For a more in-depth discussion of this topic, see Chapter Seven.) However, we do not have direct survey evidence or other data to directly assess the impact on the U.S. military.

Evidence from the General U.S. Population

According to recent research on the U.S. general population, attitudes toward transgender individuals are significantly more negative than attitudes toward other sexual minorities (Norton and Herek, 2013). However, heterosexual adults' positive attitudes toward and acceptance of transgender individuals are strongly correlated with their attitudes and acceptance of gay, lesbian, and bisexual individuals (Flores, 2015). As such, similar to changes seen in public attitudes toward homosexuality, tolerance and acceptance toward the transgender population could change over time. Additionally, evidence does indicate that direct interactions with transgender individuals significantly reduce negative perceptions and increase acceptance (Flores, 2015), which would suggest that those who have previously interacted with transgender individuals would be more likely to be tolerant and accepting of them in the future. Similar findings have arisen from surveys and focus groups with service members regarding attitudes toward the integration of women into direct combat positions (Szayna et al., 2015) and attitudes toward allowing gay and lesbian service members to serve openly in the U.S. military (RAND National Defense Research Institute, 2010).[7]

---

[7]   A recent article examined the attitudes of military academy, Reserve Officers' Training Corps, and civilian undergraduates in the United States toward transgender people in general, in the workplace, and in the military (see Ender, Rohall, and Matthews, 2016).

Evidence from Foreign Militaries

While there are limited data on the effects of transgender personnel serving openly in foreign militaries, the available research revealed no significant effect on cohesion, operational effectiveness, or readiness. In the case of Australia, there is no evidence and there have been no reports of any effect on cohesion, operational effectiveness, or readiness (Frank, 2010). In the case of Israel, there has also been no reported effect on cohesion or readiness (Speckhard and Paz, 2014). Transgender personnel in these militaries have reported feeling supported and accommodated throughout their gender transition, and there is no evidence of any impact on operational effectiveness (Speckhard and Paz, 2014). In fact, commanders have reported that transgender personnel perform their military duties and contribute effectively to their units (Speckhard and Paz, 2014). Interviews with commanders in the United Kingdom also found no effect on operational effectiveness or readiness (Frank, 2010). Some commanders reported that increases in diversity had led to increases in readiness and performance. Interviews with these same commanders also found no effect on cohesion, though there were some reports of resistance to the policy change within the general military population, which led to a less-than-welcoming environment for transgender personnel. However, this resistance was apparently short-lived (Frank, 2010).

The most extensive research on the potential effects of openly serving transgender personnel on readiness and cohesion has been conducted in Canada. This research involved an extensive review of internal defense reports and memos, an analysis of existing literature, and interviews with military commanders. It found no evidence of any effect on operational effectiveness or readiness. In fact, the researchers heard from commanders that the increased diversity improved readiness by giving units the tools to address a wider variety of situations and challenges (Okros and Scott, 2015). They also found no evidence of any effect on unit or overall cohesion. However, there have been reports of bullying and hostility toward transgender personnel, and some sources have described the environment as somewhat hostile for transgender personnel (Okros and Scott, 2015).

To summarize, our review of the limited available research found no evidence from Australia, Canada, Israel, or the United Kingdom that allowing transgender personnel to serve openly has had any negative effect on operational effectiveness, cohesion, or readiness. However, it is worth noting that the four militaries considered here have had fairly low numbers of openly serving transgender personnel, and this may be a factor in the limited effect on operational readiness and cohesion.

Costs of Separation Requirements Related to Transgender Service Members

We considered the costs and benefits of providing appropriate care to transgender service members, the requirements for those who would serve openly if the current policy changed, and the costs of continuing the current administrative separation process. We analyzed the costs of separation under several assumptions: (1) some transgender personnel are currently serving but are not able to reveal their transgender status, (2) some individuals who would be desirable recruits could be excluded for reasons only related to their gender identity, and (3) some individuals who are transgender are or have been separated for reasons only related to their gender identity, which imposes separation costs.

Separation and a continued ban on open service (i.e., manpower losses) are the alternatives to meeting the medical needs of transgender individuals. As detailed in Chapter Two, the continued ban on open service may result in worsening mental health status, declining productivity, and other negative outcomes due to lack of treatment for gender identity–related issues. In addition, if DoD actively pursues separation, the process can be tedious, especially now that it requires the approval of the Under Secretary of Defense for Personnel and Readiness. Under current DoD regulations, transgender personnel can be declared administratively unfit for service if their gender identity affects their ability to meet operational or duty requirements. A June 2015 revision to DoD policy requires that a discharge justification be based on inability to meet duty requirements. However, any "administratively unfit" finding prohibits the individual from being medically evaluated for continued service.[8] Absent this process, transgender service members do not have recourse to allow mental health experts or medical professionals to review their case concurrently. This can result in unnecessary and inconsistent approaches to discharging transgender service members. As was the case in enforcing the policy on homosexual conduct, this can involve costly administrative processes and result in the discharge of personnel with valuable skills who are otherwise qualified (U.S. Government Accountability Office, 2011).

Moreover, the total cost in lost days available for deployment is negligible and significantly smaller than the lack of availability due to medical conditions. For example, in 2015 in the Army alone, there were 102,500 nondeployable soldiers, 50,000 of whom were in the AC (Tan, 2015). This accounted for about 14 percent of the AC—personnel who were ineligible to deploy for legal, medical, or administrative reasons.

---

[8]   These boards provide an established process and mechanism for evaluating whether a service member with an ailment or diagnosis, such as a mental health diagnosis, could continue military service. The services use the Medical Evaluation Board and Physical Evaluation Board systems to determine whether personnel "with an ailment or diagnosis, such as a mental health diagnosis, can continue . . . military service," based on a thorough review of fitness to serve (DoDI 1332.38, 1996).

Of those, 37,000 could not deploy due to medical conditions.[9] Excluding those who were severely injured and required longer-term care, there were 28,490 service members who had either category 1 (up to 30 days) or category 2 (more than 30 days) restrictions. Assuming those in category 1 cannot deploy for 30 days and those in category 2 cannot deploy for 90 days, we estimate there are currently 5,300 nondeployable labor-years in the Army alone. Thus, we anticipate a minimal impact on readiness from allowing transgender personnel to serve openly.

---

[9] Rates of injury and nondeployability time as reported in Cox (2015).

CHAPTER SEVEN

What Lessons Can Be Learned from Foreign Militaries That Permit Transgender Personnel to Serve Openly?

As the U.S. military considers changes to its transgender personnel policy, revisions to several other policies may be necessary. Policies in need of change would cover a range of personnel, medical, and operational issues affecting individuals and units, including some policies that currently vary by gender. Examples of the latter would include housing assignments, restrooms, uniforms, and physical standards. While these are new questions for the U.S. military, there are other countries that already allow transgender personnel to serve openly in their militaries and have already addressed these policy issues.

We reviewed policies in foreign militaries that allow transgender service members to serve openly. Our primary source for the observations presented in this report was an extensive document review that included primarily publicly available policy documents, research articles, and news sources that discussed policies on transgender personnel in these countries. The information about the policies of foreign militaries came directly from the policies of these countries as well as from research articles describing the policies and their implementation. Our findings on the effects of policy changes on readiness draw largely from research articles that have specifically examined this question using interviews and analyses of studies completed by the militaries themselves. Finally, our insights on best practices and lessons learned emerged both directly from research articles describing the evolution of policy and the experiences of foreign militaries and indirectly from commonalities in the policies and experiences across our four case studies. Recommendations provided in this report are based on these best practices and lessons learned, as well as a consideration of unique characteristics of the U.S. military.

This review and analysis of the policies in foreign militaries can serve as a reference for U.S. decisionmakers as they consider possible policy revisions to support the integration of openly transgender personnel into the U.S. military. We include information on how, when, and why each country changed its policy. We also detail the policies of each country, covering such issues as the medical and administrative requirements before gender transition can begin, housing assignments, uniform wear, and physical fitness standards.

Policies on Transgender Personnel in Foreign Militaries

According to a report by the Hague Center for Security Studies, there are 18 countries that allow transgender personnel to serve openly in their militaries: Australia, Austria, Belgium, Bolivia, Canada, Czech Republic, Denmark, Estonia, Finland, France, Germany, Israel, Netherlands, New Zealand, Norway, Spain, Sweden, and the United Kingdom (Polchar et al., 2014). This chapter describes the policies of the four countries—Australia, Canada, Israel, and the United Kingdom—with the most well-developed and publicly available policies on transgender military personnel. It focuses explicitly on policies that describe how these foreign militaries treat transgender personnel and how they address this population's gender transition needs. While the focus of the chapter is on the specific policies integrating openly transgender military personnel in these four foreign militaries, we also provide some information about what happened after the policy change, including bullying and harassment, and summarize best practices and challenges that emerged from our four case studies.[1]

The formal policies on transgender personnel in the four countries address a number of aspects of the gender transition process.[2] Generally, these policies do not explicitly address such issues as the recruitment or retention of transgender personnel, though we provide information on the qualification of transgender personnel to serve when it is available. They do generally address such issues as the requirements for transitioning, housing assignments, restroom use, uniforms, identity cards, and physical standards. They also address whether the transitioning personnel remain with their old units or shift to new ones and how other members of a unit should be informed. Finally, the policies address access to medical care and what is or is not covered by the military health care system.

In addition to addressing these crucial issues, foreign military policies on transgender personnel typically lay out a gender transition plan, which describes the timeline or steps in the transition process. However, it is worth noting that each individual's

---

[1]   We looked for information on the policies of the other 14 countries but were unable to find any publicly available documents in English.

[2]   We note a few interesting points about other countries that we investigated but for which we were unable to find sufficient publicly available information to construct a complete case. The Netherlands was the first country to allow transgender personnel to serve openly in its military, opening its ranks in 1974. New Zealand opened its military to transgender personnel in 1993; although we could not find a written policy, a 2014 report by Hague Center for Strategic Studies referred to New Zealand's as the most friendly military to transgender personnel. The New Zealand Defence Force also has an advocacy group, OverWatch, that provides support to lesbian, gay, bisexual, and transgender personnel (see Polchar et al., 2014).

gender transition is unique. While some choose to undergo hormone therapy or gender reassignment surgery, this is not required for gender transition. As a result, the timelines outlined in the policies are intended to be examples only.

Australia

In 2010, the Australian Defence Force revoked the defense instruction that prohibited transgender individuals from serving openly, stating that excluding transgender personnel from service was discrimination that could no longer be tolerated (Ross, 2014). The Australian Department of Defence, with the advocacy group Defence Lesbian, Gay, Bisexual, Transgender, and Intersex Information Service, has produced guides to support commanders, transitioning service members, and the units in which transitioning members are serving (Royal Australian Air Force, 2015). The guide outlines five stages in the gender transition process: diagnosis, commencement of treatment, disclosure to commanders and colleagues, the post-transition experience, and, if applicable, gender reassignment surgery (Royal Australian Air Force, 2015). There is no public information on the number of transgender personnel in the Australian military or the costs associated with covering gender transition–related medical care.

A service member's gender transition begins after receiving a medical diagnosis of gender incongruence from a doctor approved by the Australian Defence Force. According to Australian Defence Force policy, once service members receive this diagnosis and present a medical certification form to their commanders, they can begin the "social transition," which policy defines as the time when an individual begins living publicly as the target gender. Under the current policy, after this point, the service member's administrative record is updated to indicate the target gender for the purposes of uniforms, housing, name, identification cards, showers, and restrooms (Royal Australian Air Force, 2015). This means that, after this point, the service member is assigned to housing of the target gender, may use the restrooms of the target gender, has an identification card with the target gender and new name, and can wear the uniform of the target gender.

During the social transition, the service member may undergo hormone therapy. However, neither hormone therapy nor gender reassignment surgery is required for the administrative changes to occur. Importantly, this shift in gender for military administrative purposes may not always match the legal transition (with respect to the Australian government) to the target gender (Royal Australian Air Force, 2015). Finally, when transgender service members choose to transition, they may choose whether to stay with their current unit or transfer to a different one. They may also choose how colleagues are informed of the gender transition—that is, whether they wish to tell colleagues themselves or have a senior leader do so.

Australia's policy also addresses matters related to physical standards and medical readiness. During the transition period, a service member may be downgraded in terms of physical readiness or declared unable to deploy for some time. However, this determination is decided on a person-by-person basis and is only temporary. According to the guide provided to service members and commanders, most individuals are placed on "MEC [Medical Employment Classification] 3—Rehabilitation" status during their medical transition or if they require four consecutive weeks of sick leave. Others may be able to remain "MEC 2—Employable and Deployable with Restrictions" for the majority of the gender transition period. In most cases, this determination is made by a certification board, though commanders are also given discretion to downgrade transitioning service members or declare them unfit to deploy, contingent on a stated inability to accommodate the service member's needs or a determination that the transitioning service member's presence would undermine the unit's performance. However, there is no public information available on the types of justifications a commander might give in making such a determination.

The deployment status of each individual will vary during the gender transition based on the transition path chosen (for example, whether hormone therapy or surgery is undertaken). Some of these treatments are covered by military health care. In Australia, medical treatments associated with gender transition, including both hormone therapy and gender reassignment surgery, are covered, but treatments considered "cosmetic" might not be (Royal Australian Air Force, 2015). However, it is not clear what is classified as cosmetic or what might be considered medically necessary. Importantly, gender transition–related medical procedures are provided only at certain facilities, so service members who wish to receive these treatments may need to make special requests for specific assignments where their needs can be met. In general, personnel are permitted to take sick leave to facilitate their medical transition (Royal Australian Air Force, 2015).

Transitioning service members' deployment status will also depend on their ability to meet physical fitness standards. During the transition period, a service member may be considered medically exempt from meeting physical fitness standards, with a coinciding readiness classification of nondeployable. Once deemed medically able to complete the test by a medical professional, the service member may be asked to meet the standards of the target gender. However, which gender standards the individual is required to meet and when is determined by the medical officer overseeing the gender transition (Royal Australian Air Force, 2015). Thus, the point at which each transitioning service member is required to meet the target-gender standards varies.

Canada

In Canada, a 1992 lawsuit from a member of the armed forces resulted in the repeal of a regulation banning gay, lesbian, and transgender individuals from serving openly in the military (Okros and Scott, 2015). In 1998, the Canadian military explicitly recognized gender identity disorder and agreed to cover gender reassignment surgery. In 2010, Canadian military policy was revised to clarify transgender personnel issues, such as name changes, uniforms, fitness standards, identity cards, and records (Okros and Scott, 2015). An updated policy, Military Personnel Instruction 01/11, "Management of Transsexual Members," was released in 2012 (Canadian Armed Forces, 2012). It stated, "The CF [Canadian Forces] shall accommodate the needs of CF transsexual members except where the accommodation would: constitute undue hardship; or cause the CF member to not meet, or to not be capable of meeting. . . . Minimum Operational Standards Relating to Universality of Service" (Canadian Armed Forces, 2012, p. 5). Other considerations that can be used to determine whether an accommodation is reasonable include cost and the safety of other service members and the public (Canadian Armed Forces, 2012, p. 5). Data suggest that there are approximately 265 transgender personnel serving openly and that the Canadian military pays for about one gender reassignment surgery per year (Okros and Scott, 2015).

Canada's policy on transgender personnel covers such issues as housing, identification cards, restrooms, physical standards, deployment, medical treatment, and uniforms. The process is similar in most ways to that in Australia, described earlier. In Canada, one of the first steps in the gender transition process is a medical assessment in which the individual is given a diagnosis of gender incongruence and assigned a temporary medical category that defines both employment limitations and accommodations that will be needed to support the service member during gender transition. After receiving this diagnosis, service members are responsible for informing their commanders and are asked to give commanders as much notice as possible before beginning their gender transition. After that, the service member, the service member's manager, and the unit's commanding officer are expected to meet to discuss the service member's gender transition plan and to addresses any necessary accommodations. The policy recommends frequent meetings between the service member and relevant leaders and medical professionals to ensure that the transitioning service member's needs are met. The policy also identifies subject-matter experts, such as chaplains and mental health professionals, who might be available to provide advice (Canadian Armed Forces, 2012).

The policy states that the gender transition plan should address housing, uniforms, deployments, and other administrative considerations. While the timeline will vary for each individual, in most cases, after receiving the diagnosis and informing the commander, the service member is able to begin living openly as the target gender. At this point, the service member is assigned to housing of the target gender, given ID cards with the target gender and new name, given uniforms of the target gender, and permitted to use restrooms of the target gender. However, while the individual is considered a member of the target gender for all administrative purposes within the military at this point, an official name and gender change in the military personnel system requires both medical certificates and legal documentation (Canadian Armed Forces,

2012).[3] Finally, medals and awards earned by the service member prior to transitioning cannot be transferred to the new name when the service member transitions to the target gender (Okros and Scott, 2015).

While the policy expects accommodations to be made to meet the needs of transgender personnel, it also notes that commanders must strike a balance between meeting the needs and legal rights of transgender personnel and the privacy needs of other service members in restrooms, showers, and housing. It does not, however, provide guidance on how this should be accomplished (Canadian Armed Forces, 2012). The policy also makes clear that incidents of harassment must be dealt with according to the Canadian military's discrimination and harassment policy. Finally, if the transgender service member is assigned to a new unit permanently or temporarily, any required accommodations are to be communicated to the new commanding officer prior to the service member's arrival (Canadian Armed Forces, 2012).

The medical assessment and gender transition plan developed at the start of transition are also used to determine a service member's readiness status and deployability. The policy states that service members can be downgraded temporarily in terms of their readiness, ability to deploy, and eligibility for remote assignments until gender transition is complete (Canadian Armed Forces, 2012). This determination is made primarily by the medical professionals overseeing the service member's gender transition. After the gender transition is complete, the continued need for a reduced medical standard is decided on a case-by-case basis based on the service member's overall health, chronic conditions, and need for access to medical care. After beginning the gender transition, and based on the medical assessment, the service member is considered medically exempt from physical fitness testing and requirements until legally assuming the acquired or target gender (which, as noted earlier, requires provincial recognition). At that point, the fitness standards for the acquired or target gender apply. More specifically, once personnel are removed from the medical exemption list, they have 90 days to meet the new standards (Canadian Armed Forces, 2012).

A reduced medical readiness determination during gender transition is intended primarily to ensure that the service member has uninterrupted access to medical care. Once gender transition is complete, transgender service members and their commanders responsible for identifying the service member's specific needs and how they will be addressed (Canadian Armed Forces, 2012). Gender reassignment surgery will not, however, automatically result in permanent deployment restrictions. As in Australia, gender reassignment surgery and hormone therapy are covered by military health care. The Canadian military paid for one gender reassignment surgery in 1998 and has paid for one or two surgeries per year since then (Canadian Armed Forces, 2012).

---

[3] Also note that the requirements for the legal change vary by province but typically involve only a statement that the individual has assumed the target gender and a medical certification from a doctor of a diagnosis of gender incongruence.

Israel

The Israel Defense Forces (IDF) have allowed transgender personnel to serve openly since 1998 (Speckhard and Paz, 2014).[4] The IDF experience with transgender personnel is somewhat unique because Israel's military is composed largely of conscripts who serve two or three years and then serve in the reserves with extended periods of active service. As a result, a very high percentage of the population spends extended periods of time mixing military and civilian life. From the perspective of this report, this blending of civilian and military life creates unique challenges for transgender personnel, as they cannot be one person in their civilian life and then a different person in their military life. Some transgender individuals receive a discharge or exemption from their military service based on their gender incongruence, but this decision is currently at the discretion of the commander. There is no official IDF policy on transgender personnel, but according to one report, senior members of the IDF are working to draft one (Speckhard and Paz, 2014). In 2014, the IDF announced that it would support transgender individuals throughout the transition process. Under this new policy, transgender teens who have not yet begun to transition to another gender will be enlisted according to their birth sex, but after enlistment, they will be given support and assistance with the gender transition process (Zitun, 2014). As a result, Speckhard and Paz (2014) noted, experiences vary for transgender personnel in the IDF. Some individuals report that once they ask to transition, they are allowed to dress and serve as their target gender. However, it is unclear how generalizable this is.

Typically, IDF administrative records use the gender at that time of enlistment. Since conscription occurs at age 18, and because hormone treatment for gender incongruence cannot legally begin until age 18, the administrative records of most personnel show their birth gender. Under a newly announced policy, personnel enlisted using their birth gender who identify as transgender can immediately receive support and treatment to begin the gender transition (Zitun, 2014). Importantly, however, as of 2014, the military identification card carries the birth gender until a service member undergoes gender reassignment surgery, even if the service member is living publically as the target gender (Speckhard and Paz, 2014). It should be noted that, in Israel, only one hospital can perform gender reassignment surgery, and this surgery cannot be performed until age 21, though some people go abroad for it (Speckhard and Paz, 2014). This creates some complications for housing and other matters, discussed in more detail later. The new policy will also allow transgender recruits to receive support for gender transition after enlistment.

Available evidence suggests that, in the IDF, assignment of housing, restrooms, and showers is typically linked to the birth gender, which does not change in the military system until after gender reassignment surgery. Service members who are undergo-

---

[4]   We do not know the exact date for this change because there was never a formal policy allowing or prohibiting transgender personnel from serving. It was in 1998 that the first openly transgender individual served in the IDF.

ing gender transition are accommodated, however, through the use of ad hoc solutions, including giving transitioning personnel their own showers, housing, or restrooms (Speckhard and Paz, 2014). Once transitioning personnel have completed gender reassignment surgery, they can be assigned to the housing, restrooms, and showers of their acquired gender. It is also worth noting that the majority of noncombat personnel are able to live at home, off base. As a result, the housing issue does not affect a large number of transitioning personnel (Speckhard and Paz, 2014). The issue of uniforms is usually easier to address, and service members are able to wear the uniform of the target gender once they begin their gender transition.

In addition to addressing housing and other administrative matters for conscripts and career soldiers, the IDF must address transitioning reservists. The limited information available suggests that the approach to addressing the needs of this group also varies from person to person. Usually, if reserve members are in the process of transitioning or have transitioned when called to active duty, they are permitted to return to service as their target or acquired gender (following the same administrative policies described earlier). For example, a service member who served in an all-male combat unit and is transitioning to female may be moved to another position. Again, many reservists serve their duty while living at home, so housing is not usually an issue. Restroom and shower assignments are addressed on an ad hoc basis (Speckhard and Paz, 2014). Finally, some personnel who have transitioned or are in the process of transitioning are exempted from their reserve duty. However, this is becoming less common as the IDF strives to accommodate the needs of these personnel rather than exempting them from service (Speckhard and Paz, 2014).

The IDF does not have a formal policy on physical standards for transgender individuals serving their conscription duty, reserve duty, or as professional soldiers. Available information suggests only that transgender personnel can serve in any unit or occupation for which they meet the requirements, with the exception of a few male-only combat units and certain security-related positions (Speckhard and Paz, 2014). Personnel transitioning from female to male are able to serve in male-only combat units only if they can meet the requirements set for other men. Personnel transitioning from male to female cannot serve in male-only combat units once they begin hormone treatment (Speckhard and Paz, 2014).

There do appear to be some limitations on the assignment of transgender personnel, particularly in combat units. Because of austere living conditions in these types of units, necessary accommodations may not be available for service members in the midst of a gender transition. As a result, transitioning individuals are typically not assigned to combat units (Speckhard and Paz, 2014). Transgender personnel are also limited from assignment to certain security-related positions due to concerns about blackmail, based on the assumption that these service members might be open about their gender identity in the military but might not have told others, including family members. Keeping these types of secrets might make an individual susceptible to blackmail or extortion (Speckhard and Paz, 2014).

In the IDF, medical issues and matters related to the readiness of transgender personnel are addressed on a case-by-case basis, though a more formal policy is being developed. For conscripts, the only treatment that can be provided by the military is hormone therapy because gender reassignment surgery is possible in Israel only after age 21, by which point the conscription duty is usually completed (Speckhard and Paz, 2014). Those who choose to stay in the military full-time after the age of 21, as well as those in the reserve called to back to active service, may receive both hormone therapy and gender reassignment surgery. Those who choose to undergo surgery are permitted to take a period of sick leave for the surgery and recovery, as they can for any other medical treatment or surgery (Speckhard and Paz, 2014). Israel has nationalized health care that typically covers all treatments associated with gender transition, ranging from psychiatric care to pre- and postoperative care, hormone treatment, breast augmentation, and facial feminization. Apart from the approaches used to address physical standards for transitioning individuals (discussed earlier), there are no specific policies governing the readiness classification of transitioning IDF personnel, though some are in development (Zitun, 2014).

United Kingdom

The United Kingdom lifted the ban on transgender personnel in 2000 following a European Court of Human Rights ruling that the country's policy violated the right to privacy under the European Convention on Human Rights (Frank, 2010). The policy change was implemented with guidance to commanders, as well as a code of social conduct that allowed commanders to address inappropriate behavior toward transgender personnel by appealing to broader principles of tolerance and diversity and to guard operational effectiveness (Yerke and Mitchell, 2013). In 2009, the British Armed Forces released the "Policy for Recruitment and Management of Transsexual Personnel in the Armed Forces" to offer clearer guidance to commanders on how gender transition–related issues should be addressed (Yerke and Mitchell, 2013). While transgender personnel are able to serve openly, under the current policy, they can be excluded from sports that organize around gender to ensure the safety of the individual or other participants. The British Army also provides its official policy on transgender personnel on its website:

> The Army welcomes transgender personnel and ensures that all who apply to join are considered for service subject to meeting the same mental and physical entry standard as any other candidate. If you have completed transition you will be treated as an individual of your acquired gender. Transgender soldiers serve throughout the Army playing their part in the country's security. There is a formal network that operates in the Army to ensure that transgender soldiers can find advice and support with issues that affect their daily lives. (British Army, undated)

However, the military encourages those who have not yet started their gender transition to complete their transition before joining (UK Ministry of Defence, 2009).

The 2009 UK policy is similar to those in Canada and Australia in terms of the areas covered and approaches to addressing key issues, though the UK policy provides some additional room for individual differences. The policy also includes an extensive discussion of the legal and privacy protections afforded to transgender personnel. These protections are important because they also apply to administrative and medical records in the military system.

The UK policy defines five stages of gender transition: diagnosis, social transition (the individual begins living openly as the target gender), medical treatment/hormone therapy, surgical reassignment, and postoperative transition. However, it also recognizes that the process of gender transition may be different for each person. The policy suggests that each individual work with commanders and service authorities to develop a plan that includes a timeline for transition. The gender transition plan agreed to by the service member and commanders should specify the timing of changes, such as to housing assignments and uniforms. The specific point at which a service member transitions for the purposes of name, uniform, housing, restrooms, and ID cards may vary from person to person. Typically, when service members begin living publicly as the target gender (the social transition) they are reassigned to housing of the target gender, use the restrooms and uniforms of the target gender, and are given an ID card indicating that they are a member of the target gender. Importantly, this shift in gender for administrative purposes does not have to correspond to the point at which an individual transitions gender within the UK legal system, a process that involves a diagnosis of gender incongruence and two years of living as the acquired gender (UK Ministry of Defence, 2009). The policy also notes that it is unlawful to force transgender personnel to use separate toilet or shower facilities or occupy separate housing accommodations from the rest of the force.

The gender transition plan addresses other logistics of the transition. For example, it should specify scheduled time off required for medical procedures, including gender reassignment surgery. In general, medical treatment associated with gender transition is treated like any other medical issue experienced by a service member. However, while hormone replacement therapy is covered by military health care, gender reassignment surgery is not (UK Ministry of Defence, 2009). The policy notes that the timeline and timing of the transition must take into consideration the needs of the service. As a result, at least four weeks notice is typically needed prior to the start of a service member's gender transition. The gender transition plan should also specify whether service members wish to transition in their current post or transfer to a new position and whether they want to tell their colleagues about the gender transition themselves or would like someone else to do this. This decision may depend on the size of the unit. In a small unit, it may be easy to inform fellow service members personally. In a larger organization, it may not be necessary to tell every individual. Commanders of units with transgender personnel are encouraged to consult members of the Service Equality and Diversity staff about how to approach education and management in matters associated with transgender service members.

The UK policy also addresses medical readiness and physical standards. Transgender personnel are evaluated for medical readiness and deployability on a case-by-case basis following a medical evaluation. During the transition period, specifically during hormone treatment and immediately before and after surgery, service members may receive a reduced Medical Employment Standard, which restricts deployability and sea service (UK Ministry of Defence, 2009). Transitioning service members who continue to meet physical standards throughout this period and are able to perform their jobs may retain normal readiness standards. Usually, those who do not undergo hormone therapy or gender reassignment surgery are able to maintain a fully deployable status throughout their gender transition (UK Ministry of Defence, 2009). Service members who are undergoing hormone therapy are able to deploy, as long as the hormone dose is steady and there are no major side effects. However, deployment to all areas may not be possible, depending on the needs associated with any medication (e.g., refrigeration). Some service members may also be required to have a psychiatric evaluation, but only if they show signs of mental health distress (UK Ministry of Defence, 2009). Individuals who have finished their gender transition and can meet the requirements of their legal gender are considered fully deployable. However, those who remain in a state of reduced readiness for an extended period may have to be discharged (UK Ministry of Defence, 2009). Importantly, the British military encourages individuals who are in the midst of their gender transition and are considering joining the military to wait until the gender transition is complete before joining, as the military may not always be able to provide the support the individual needs during gender transition.

The specific physical standards a transitioning individual must meet during and after the gender transition period are determined on a case-by-case basis. The policy allows that there may be a period of time—especially for individuals transitioning from female to male—during which a service member is not yet able to meet the standards of the target gender. In these cases, medical staff and commanders may assess the individual and determine the appropriate interim standards (UK Ministry of Defence, 2009). Once the gender transition is considered "complete," personnel are required to meet the standards of the target gender (UK Ministry of Defence, 2009). However, the policy recognizes that the point at which the gender transition is complete may vary: It may be complete after hormone therapy or after surgery, or simply after the individual begins living as the target gender. Therefore, the policy continues to allow for some flexibility in physical standards, even for members at the end of their gender transition process (UK Ministry of Defence, 2009). Modified standards may be set by medical staff and commanders, if necessary. Continued failure to meet whatever physical standards are determined to be appropriate (modified or otherwise) can lead to administrative discharge (UK Ministry of Defence, 2009).

The policy also addresses positions that are "gender-restricted" or have unique standards. The United Kingdom still has a number of combat occupations closed to women. Personnel who are transitioning from male to female may not serve in male-only occupations as long as this policy remains in place. Those transitioning from female to male may hold these jobs, assuming that they are able to meet the physical standards (UK Ministry of Defence, 2009). Transgender personnel may hold positions that have unique standards related to the occupation, as long as they can meet the physical and other requirements for the specific position. Finally, according to the policy, service members may request that their medals be transferred to a new name by submitting the request in writing. They are allowed to continue wearing qualifications earned while serving as their birth gender. However, this may indicate their transgender status to others (UK Ministry of Defence, 2009).

Effects on Cohesion and Readiness

As indicated in Chapter Six, while there is limited research on the effects of transgender personnel serving openly in foreign militaries, the available evidence indicated no significant effect on cohesion, operational effectiveness, or readiness. In the Australian case, there is no evidence and there have been no reports of any effect on cohesion, operational effectiveness, or readiness (Frank, 2010). In the Israeli case, there has also been no reported effect on cohesion or readiness (Speckhard and Paz, 2014). Transgender personnel in these militaries report feeling supported and accommodated throughout their gender transition, and there has been no evidence of any effect on operational effectiveness (Speckhard and Paz, 2014). As noted earlier, commanders report that transgender personnel perform their military duties and contribute to their units effectively (Speckhard and Paz, 2014). Interviews with commanders in the United Kingdom also found no effect on operational effectiveness or readiness (Frank, 2010). Some commanders reported that increases in diversity had led to increases in readiness and performance. Interviews with these same commanders also found no effect on cohesion, though there were some reports of resistance to the policy change within the general military population, which led to a less-than-welcoming environment for transgender personnel. However, this resistance was apparently short-lived (Frank, 2010).

The most extensive research on the potential effects of openly serving transgender personnel on readiness and cohesion has been conducted in Canada. This research involved an extensive review of internal defense reports and memos, an analysis of existing literature, and interviews with military commanders. It found no evidence of any effect on operational effectiveness or readiness. In fact, the researchers heard from commanders that the increased diversity improved readiness by giving units the tools to address a wider variety of situations and challenges (Okros and Scott, 2015). They also found no evidence of any effect on unit or overall cohesion. However, there have been reports of bullying and hostility toward transgender personnel, and some sources have described the environment as somewhat hostile for transgender personnel (Okros and Scott, 2015).

To summarize, our review of the limited available research found no evidence from Australia, Canada, Israel, or the United Kingdom that allowing transgender personnel to serve openly has had any negative effect on operational effectiveness, cohesion, or readiness. However, it is worth noting that the four militaries considered here have had fairly low numbers of openly serving transgender personnel, and this may be a factor in the limited effect on operational readiness and cohesion.

Best Practices from Foreign Militaries

Several best practices and lessons learned emerged both directly from research articles describing the evolution of policy and the experiences of foreign militaries and indirectly from commonalities in the policies and experiences across our four case studies. The best practices that extended across all cases include the following:

The Importance of Leadership

Sources from each of our case-study countries stressed that leadership support was important to executing the policy change. Leaders provided the impetus to draft and implement new policies and were integral to communicating a message of inclusion to the entire force. Supportive leaders were also important in holding accountable those personnel who participated in discrimination (Okros and Scott, 2015; Speckhard and Paz, 2014). Each of the cases underscores the importance of having strong leadership support to back and enforce the policy change, along with clearly written policies that are linked to national policy wherever possible (Frank, 2010). The militaries found that presenting a "business case" for diversity and emphasizing the advantages of an inclusive military, including better retention and recruiting, can help reduce resistance to a policy change (Frank, 2010).

Awareness Through Broad Diversity Training

The most effective way to educate the force on matters related to transgender personnel is to integrate training on these matters into the diversity and harassment training already given to the entire force. This training addresses all forms of harassment and bullying, including that based on religion, race, and ethnicity (Frank, 2010; Okros and Scott, 2015; Belkin and McNichol, 2000–2001).

In the four cases we reviewed in-depth, we found that targeting only commanders with training and information on what it means to be transgender is not as effective in fostering an inclusive and supportive environment as training that targets the entire force and is integrated into broader forcewide diversity training. The foreign militaries that we examined train not only units with transitioning individuals but also the entire force by including gender identity alongside sexual orientation, religion, ethnicity, and other markers of difference in diversity training and education. However, efforts must be made simultaneously to protect the privacy of transitioning service members. In some cases, telling a unit that a transgender member is arriving before that individual arrives can be counterproductive (Frank, 2010).

The Importance of an Inclusive Environment

An all-inclusive military environment—not just as it pertains to transgender personnel, sexual orientation, or gender identity, but a culture that embraces diversity—can support the integration of openly serving transgender personnel. In this context, gender identity is just one marker of diversity.[5]

Ensuring Availability of Subject-Matter Experts to Advise Commanders

Most of the four countries we examined in-depth also make subject-matter experts (e.g., chaplains, psychiatrists) and gender advisers (individuals who have special training in gender awareness and gender mainstreaming in the military context) available to commanders tasked with the integration of transgender personnel. Gender advisers were originally intended to deal primarily with issues associated with integrating women into male-dominated military environments, but they could also help with other gender-related matters, including transgender personnel policy. They serve directly within military units and are a readily available resource to commanders. Adopting a similar practice of integrating advisers with expertise in the area of transgender personnel policy and gender transition-related matters might also support the integration of transgender service members in the U.S. military.

Lessons Learned and Issues to Consider for U.S. Military Policy

Based on these best practices and the broader experiences of four foreign militaries, there are some key lessons to be learned and possible issues to consider when crafting U.S. military transgender personnel policy. First, in each of the four foreign militaries, there were some reports of resistance, bullying, and harassment of transgender personnel who made their gender transition public. This harassment ranged from exclusion to more aggressive behavior. In most cases, this behavior was relatively limited; however,

---

[5]   Remarks by a Canadian subject-matter expert in a phone discussion with RAND researchers, November 2015.

in some cases, it did contribute to a hostile work environment for transgender personnel and had the effect of discouraging these personnel from being open about their gender transition or gender identity (Okros and Scott, 2015; Frank, 2010). Although the foreign militaries we examined tended to adopt a policy of no tolerance for this type of harassment, some bullying behavior may have gone unreported (Okros and Scott, 2015; Frank, 2010). In the case of Canada, the issue of restrooms for transgender personnel is an ongoing topic of discussion, and restrooms have been a common site of harassment and discrimination (Okros and Scott, 2015).

A second lesson learned is related to problems caused by the lack of an explicit, clearly written policy. For instance, in the IDF, without a clear policy, some transitioning individuals are placed in difficult and uncomfortable situations. For example, in some cases, personnel who have been permitted to begin hormone therapy cannot be housed with members of their target gender or grow their hair and fingernails (in the case of individuals transitioning from male to female). Others have been isolated, assigned to separate housing, or asked to use separate restrooms (Speckhard and Paz, 2014). Recognizing these challenges, IDF leadership is working to design a clear and explicit policy. In the Israeli case, transgender individuals were allowed to serve openly before a formal policy was written. Only when it was faced with questions about the integration of transgender personnel did the IDF begin to create a formal policy.[6] In Canada, a similar policy gap arose when transgender personnel were allowed to serve openly following a national policy revision that ended discrimination based on sexual orientation or gender. However, the focus at that point was on gay and lesbian service members, and no formal policy was created to address transgender personnel explicitly. When matters related to the medical care of transgender personnel arose, Canadian defense leaders developed a policy that just addressed this narrow, pressing issue, and did not develop policies to address the other matters (e.g., housing, restrooms, name changes). Commanders complained that the original policy was too vague and lacked sufficient details. A new, revised policy was written in 2012, and commanders have responded with positive feedback.[7] The lack of a clear, written policy has also been an issue in Australia.

A third and final issue that has come up in at least two of the countries we surveyed is that of awards and medals. In the UK case, medals and awards received prior to gender transition can be transferred to the service member's post-transition name (UK Ministry of Defence, 2009). In the Canadian case, this is not possible, and the awards remain associated only with the original name. This is a cause for concern among transgender personnel in the Canadian military, but Canadian officials have responded that they cannot rewrite history (Okros and Scott, 2015). This is a policy area that the United States should consider alongside other administrative policies.

---

[6] Remarks by a Canadian subject-matter expert in a phone discussion with RAND researchers, November 2015.

[7] Remarks by a Canadian subject-matter expert in a phone discussion with RAND researchers, November 2015.

CHAPTER EIGHT

Which DoD Policies Would Need to Be Changed if Transgender Service Members Are Allowed to Serve Openly?

This chapter reviews DoD accession, retention, separation, and deployment policies and provides an assessment of the impact of changes required to allow transgender personnel to serve openly. For our analysis of DoD policies, we reviewed 20 current accession, retention, separation, and deployment regulations across the services and the Office of the Secretary of Defense. We also reviewed 16 other regulations that have been replaced by more recent regulations or did not mention transgender policies.[1] DoDI 6130.03 establishes medical standards for entry into military service, including a list of disqualifying physical and mental conditions, some of which are transgender-related.[2] Current DoD policy also authorizes, but no longer requires, the discharge of transgender personnel for reasons related to both medical conditions that generate disabilities, as well as mental health concerns.[3] However, a July 2015 directive from the Office of the Secretary of Defense elevated decisions to administratively separate transgender service members to the Office of the Under Secretary of Defense for Personnel and Readiness (DoD, 2015b).

Note that our review focused on transgender-specific DoD instructions that may contain unnecessarily restrictive conditions and reflect outdated terminology and assessment processes. However, in simply removing these restrictions, DoD could inadvertently affect overall standards. While we focus on reforms to specific instruc-

---

[1] These additional policies are listed in Appendix D.

[2] The instruction specifies conditions that disqualify accessions, including "current or history of psychosexual conditions, including but not limited to transsexualism, exhibitionism, transvestism, voyeurism, and other paraphilias"; "history of major abnormalities or defects of the genitalia including but not limited to change of sex, hermaphroditism, pseudohermaphroditism, or pure gonadal dysgenesis"; and "history of major abnormalities or defects of the genitalia such as change of sex, hermaphroditism, pseudohermaphroditism, or pure gonadal dysgenesis" (DoDI 6130.03, 2011, enclosure 4).

[3] "Sexual gender and identity disorders" are specified as medical conditions that may generate disabilities under DoDI 1332.38, enclosure 5 (2006). Mental health conditions are specified in DoDI 1332.14 (2014) and DoDI 1332.30 (2013) for enlisted and officers, respectively. DoDI 1332.18, issued on August 5, 2014, updated these guidelines and established general criteria for referral for disability evaluation and defers to service-specific standards for retention. However, a recent review of this revision suggests that service-specific regulations may still disqualify transgender personnel, and the new guidance may not overrule those service policies (Pollock and Minter, 2014).

tions and directives, we note that DoD may wish to conduct a more expansive review of personnel policies to ensure that individuals who join and remain in service can perform at the desired level, regardless of gender identity.

Accession Policy

The language pertaining to transgender individuals in accession instructions does not match that used in DSM-5.[4] This results in restrictions in DoD policy that do not match current medical understanding of gender identity issues and thus may be misapplied or difficult to interpret in the context of current medical treatments and diagnoses. Under current guidelines, otherwise qualified individuals could be excluded for conditions that are unlikely to affect their military service, and individuals with true restrictions may be more difficult to screen for and identify. Modernizing the terminology to match current psychological and medical understanding of gender identity would help ensure that existing procedures do not inadvertently exclude otherwise qualified individuals who might want to join the military. We recommend that DoD review and revise the language to match the DSM-5 for conditions related to mental fitness so that mental health screening language matches current disorders and facilitates appropriate screening and review processes for disorders that may affect fitness for duty. Similarly, physical fitness standards should specify physical requirements, rather than physical conditions. Finally, the physical fitness language should clarify when in the transition process the service member's target gender requirements will begin to apply.

Retention Policy

We recommend that DoD expand and enhance its guidance and directives to clarify and adjust, where necessary, standards for retention of service members during and after gender transition. Evidence from Canada and Australia suggests that transgender personnel may need to be held medically exempt from physical fitness testing and requirements during transition (Canadian Armed Forces, 2012; Royal Australian Air Force, 2015). However, after completing transition, the service member could be required to meet the standards of the acquired gender. The determination of when the service member is "medically ready" to complete the physical fitness test occurs on a case-by-case basis and is typically made by the unit commander.

---

[4] Two key changes are that the term *transsexualism* has been replaced, and *gender dysphoria* is no longer in the chapter "Sexual Desire Disorders, Sexual Dysfunctions, and Paraphilias" but, rather, has its own chapter (Milhiser, 2014).

Separation Policy

DoD may wish to revise the current separation process based on lessons learned from the repeal of Don't Ask, Don't Tell. The current process relies on administrative decisions outside the purview of the standard medical and physical review process. This limits the available documentation and opportunities for review, and it could prove burdensome if transgender-related discharges become subject to re-review. When medically appropriate, DoD may wish to establish guidance on when and how such discharge reviews should be handled. We also recommend that DoD develop and disseminate clear criteria for assessing whether transgender-related conditions may interfere with duty performance.

Deployment Policy

Deployment conditions vary significantly based on the unique environment of each deployment, with some deployed environments able to accommodate transgender individuals, even those who are undergoing medical treatments. Moreover, recent medical advancements can minimize the invasiveness of treatments and allow for telemedicine or other forms of remote medical care. Given medical and technological advances, DoD may wish to adjust some of its processes and deployment restrictions to minimize the impact on readiness. For example, current regulations specify that conditions requiring regular laboratory visits make service members ineligible for deployment, including all service members who are receiving hormone treatments,[5] since such treatments require laboratory monitoring every three months for the first year as hormone levels stabilize (Hembree et al., 2009; Elders et al., 2014). Such a change would require DoD to either permit more flexible monitoring strategies[6] or provide training to deployed medical personnel.[7] Similarly, the use of refrigerated medications is a disqualifying condition for deployment,[8] even though nearly all hormone therapies are available in other formats that do not require refrigeration.

---

[5]   Current regulations state that "medications that require laboratory monitoring or special assessment of a type or frequency that is not available or feasible in a deployed environment" disqualify an individual from deployment (Office of the Assistant Secretary of Defense for Health Affairs, 2013, p. 3).

[6]   Some experts suggest that alternatives, such as telehealth reviews, would address this issue for rural populations with limited access to medical care (see, for example, WPATH, 2011).

[7]   "Independent duty corpsmen, physician assistants, and nurses can supervise hormone treatment initiated by a physician" (Elders et al., 2014).

[8]   The memo issued by the Office of the Assistant Secretary of Defense for Health Affairs states, "Medications that disqualify an individual for deployment include . . . [m]edications that have special storage considerations, such as refrigeration (does not include those medications maintained at medical facilities for inpatient or emergency use)" (Office of the Assistant Secretary of Defense for Health Affairs 2013, p. 3).

CHAPTER NINE
Conclusion

By many measures, there are currently serving U.S. military personnel who are transgender. Overall, our study found that the number of U.S. transgender service members who are likely to seek transition-related care is so small that a change in policy will likely have a marginal impact on health care costs and the readiness of the force. We estimate, based on state-level surveys of transgender prevalence, that between 1,320 and 6,630 transgender personnel may be serving in the AC, and 830–4,160 may be serving in the SR. Estimates based on studies from multiple states, weighted for population and the gender distribution in the military, imply that there are around 2,450 transgender service members in the AC and 1,510 in the SR.[1]

However, only a small proportion of these service members will seek gender transition–related treatment each year. Employing utilization and cost data from the private health insurance system, we estimated the potential impact of providing this care to openly serving transgender personnel on AC health care utilization and costs. Directly applying private health insurance utilization rates to the AC military population indicated that a very small number of service members will access gender transition–related care annually. Our estimates based on private health insurance data ranged from a lower-bound estimate of 29 AC service members to an upper-bound estimate of 129 annually using care, including those seeking both surgical and other medical treatments.

Using estimates from two states and adjusting for the male/female AC distribution, we also estimate a total of 45 gender transition–related surgeries, with 50 service members initiating transition-related hormone therapy annually in the AC.[2] We estimate 30 gender transition-related surgeries and 25 service members initiating hormone therapy treatments in the SR. These are likely to be upper-bound estimates, given the nonrepresentative sample selection procedures used in the NTDS. Furthermore, the best prevalence estimates that we were able to identify were from two of the more transgender-tolerant states in the country, and the empirical evidence that trans-

---

[1] Estimates are based on FY 2014 AC and SR personnel numbers.

[2] For hormone therapy recipients, the number of treatments and recipients is the same, and these estimates can be treated as counts of individuals.

gender prevalence is higher in the military than in the general population is weak. As a point of comparison, we also compared these estimated values to mental health utilization in the AC population overall. Using data from McKibben et al. (2013), we calculated that approximately 278,517 AC service members accessed mental health care treatment in 2014, the implication being that health care for the transgender population will be a very small part of the total health care provided to AC service members across the MHS.

With respect to health care costs, actuarial estimates from the private health insurance sector indicate that covering gender transition–related care for transgender employees increased premiums by less than 1 percent. Taking a weighted average of the identified firm-level data, we estimate that covering transgender-related care for service members will increase the U.S. military's AC health care spending by only 0.038–0.054 percent. Using these baseline estimates, we estimate that MHS health care costs will increase by between $2.4 million and $8.4 million. These numbers represent only a small proportion of FY 2014 AC health care expenditures ($6.27 billion) and the FY 2014 Unified Medical Program budget ($49.3 billion). This is consistent with our estimate of relatively low AC rates of gender transition–related health care utilization in the MHS.

Similarly, when considering the impact on readiness, we found that using either the prevalence-based approach or the utilization-based approach yielded an estimate of less than 0.0015 percent of total labor-years likely to be affected by a change in policy. This is much smaller than the current lost labor-years due to medical care in the Army alone.

Even if transgender personnel serve in the military at twice the rate of their prevalence in the general population and we use the upper-bound rates of health care utilization, the total proportion of the force that is transgender and would seek treatment would be less than 0.1 percent, with fewer than 130 AC surgical cases per year even at the highest utilization rates. Given this, true usage rates from civilian case studies imply only 30 treatments in the AC, suggesting that the total number of individuals seeking treatment may be substantially smaller than 0.1 percent of the total force. Thus, we estimate the impact on readiness to be negligible.

We conclude with some general recommendations and insights based on the experiences of foreign militaries that permit transgender individuals to serve openly—specifically, Australia, Canada, Israel, and the United Kingdom. Our case studies provide some guidance that policymakers should consider as they develop policies to govern the employment of transgender personnel in the U.S. military. These cases also suggested a number of key implementation practices if a decision is made to allow transgender service members to serve openly:

- Ensure strong leadership support.
- Develop an explicit written policy on all aspects of the gender transition process.

- Provide education and training to the rest of the force on transgender personnel policy, but integrate this training with other diversity-related training and education.
- Develop and enforce a clear anti-harassment policy that addresses harassment aimed at transgender personnel alongside other forces of harassment.
- Make subject-matter experts and gender advisers serving within military units available to commanders seeking guidance or advice on gender transition-related issues.
- Identify and communicate the benefits of an inclusive and diverse workforce.

Terminology

Augmentation mammoplasty: breast augmentation involving implants or lipofilling

Buccal administration: placement of medication between the gums and cheek

Chest surgery: surgery to create a contoured, male-looking chest

Clitoroplasty: surgical creation/restoration of a clitoris

Cross-dresser: someone who dresses in the clothes of the other gender, not always on a full-time basis

Female-to-male: those assigned female sex at birth who identify as male; transgender men; transmen

Gender: an individual's gender identity, which is influenced by societal norms and expectations; public, lived role as male or female

Gender assignment: initial assignment at birth as male or female; yields "natal gender" (APA, 2013, p. 451)

Gender atypical: behaviors not typical for one's gender "in a given society and historical era" (APA, 2013, p. 451)

Gender identity: "one's inner sense of one's own gender, which may or may not match the sex assigned at birth" (Office of Personnel Management, 2015, p. 2)

Gender dysphoria: "discomfort or distress that is caused by a discrepancy between a person's gender identity and that person's sex assigned at birth (and the associated gender role and/or primary and secondary sex characteristics)" (WPATH, 2011, p. 2).

Gender nonconformity: "the extent to which a person's gender identity, role, or expression differs from the cultural norms prescribed for people of a particular sex" (WPATH, 2011, p. 5, citing Institute of Medicine definition)

Gender transition–related surgery/gender-confirming surgery/sex reassignment surgery: surgery to mitigate distress associated with gender dysphoria by aligning sex characteristics with gender identity

Genderqueer: those who "define their gender outside the construct of male or female, such as having no gender, being androgynous, or having elements of multiple genders" (Roller, Sedlak, and Draucker, 2015, p. 417)

Gluteal augmentation: buttocks augmentation involving implants or lipofilling

Hormone therapy: "the administration of exogenous endocrine agents to induce feminizing or masculinizing changes" (WPATH, 2011, p. 33)

Hysterectomy: surgery to remove the uterus

Intersex: "a general term used for a variety of conditions in which a person is born with a reproductive or sexual anatomy that doesn't seem to fit the typical definitions of female or male" (Intersex Society of North America, undated)

Labiaplasty: plastic surgery for altering or creating the labia

Lipofilling: injection of fat rather than artificial implants

Male-to-female: those assigned male sex at birth who identify as female; transgender females; transwomen

Mastectomy: surgical removal of one or both breasts

Metoidioplasty: surgically relocating a clitoris that has been enlarged by hormone therapy to a more forward position that more closely resembles that of a penis; average length is 1.5–2 inches

Oophorectomy: surgical removal of one or both ovaries

Orchiectomy: surgical removal of one or both testicles

Ovariectomy: surgical removal of one or both ovaries

Parenteral administration: intravenous injection (into a vein) or intramuscular infusion (into muscle) of medication

Penectomy: surgical removal of the penis

Phalloplasty: surgical creation/reconstruction of a penis using one of a variety of techniques including free or pedicled (attached) flap (see Rashid and Tamimy, 2013)

Primary sex characteristics: physical characteristics/sex organs directly involved in reproduction

Salpingo-oophorectomy: removal of the ovaries and fallopian tubes

Scrotoplasty: surgical creation/reconstruction of testicles; in transmen, native labia tissue is used; testicular implants can be used

Secondary sex characteristics: physical characteristics that appear at puberty and vary by sex but are not directly involved in reproduction (e.g., breasts)

Sex: a person's biological status as male or female based on chromosomes, gonads, hormones, and genitals (intersex is a rare exception)

Sexual orientation: sexual identity in relation to the gender to which someone is attracted: heterosexual, homosexual, or bisexual

Thyroid chondroplasty: removal or reduction of the Adam's apple

Transdermal administration: delivery of medication across the skin with patches

Transgender: "an umbrella term used for individuals who have sexual identity or gender expression that differs from their assigned sex at birth" (Roller, Sedlak, and Draucker, 2015, p. 417)

Transsexual: someone whose gender identity is inconsistent with their assigned sex and who desires to permanently transition their physical characteristics to match their inner sense of their own gender

Urethroplasty: surgical reconstruction or fabrication of the urethra.

Vaginectomy (colpectomy): surgical removal of all or part of the vagina

Vaginoplasty: surgical creation/reconstruction of a vagina

Vulvoplasty: surgical creation/reconstruction of the vulva

History of DSM Terminology and Diagnoses

A brief historical understanding of the evolving diagnostic nomenclature pertaining to transgender status is important to discussions of related health care. DSM-III (APA, 1980) first contained the diagnosis of transsexualism. DSM-III-R (APA, 1987) introduced gender identity disorder, non-transsexual type. In DSM-IV (APA, 1994), these two diagnoses were merged and called *gender identity disorder*. Gender identity disorder, together with the paraphilias (disorders of extreme, dangerous, or abnormal sexual desire, including transvestic fetishism, sometimes referred to as cross-dressing), constituted the DSM-IV section "Sexual and Gender Identity Disorders."

With DSM-5 (APA, 2013) came the migration from *gender identity disorder* to *gender dysphoria*. The clinical significance of the shift in DSM-5 was great: For the first time, without accompanying symptoms of distress, transgender individuals were no longer considered to have a diagnosable mental disorder. The historical parallel with homosexuality is hard to miss: In 1980, DSM-III similarly normalized the DSM-II diagnosis of homosexuality, moving instead to ego-dystonic homosexuality, a diagnosis reserved only for gay persons who felt related distress. In the next DSM iteration, DSM-III-R, all reference to homosexuality as a diagnostic term was removed. In the aftermath of depathologizing gender nonconformity, a similar move relating to transgender status appears to be underway.

As noted in this report, there is a consensus among clinicians and their professional organizations that transition-related treatment with hormones or surgery constitutes necessary health care, though there is a divide over whether it serves as "a strategy to diminish the serious suffering" of the patient or "a method to assist people in finding self-actualization" (Gijs and Brewaeys, 2007, p. 184). The conclusion that transition-related surgery "is an effective treatment for gender identity disorder in adults" is based primarily on retrospective studies of satisfaction rather than randomized controlled trials or prospective studies (Gijs and Brewaeys, 2007, p. 199). The prevalence of postoperative regret is very low, though "little empirical research has been done" on related risk and protective factors (Gijs and Brewaeys, 2007, pp. 201, 204). Overall, surgery is considered "the most appropriate treatment to alleviate the suffering of extremely gender dysphoric individuals," but rigorous controlled-outcome studies evaluating its effectiveness should be conducted despite feasibility and ethical challenges (Gijs and Brewaeys, 2007, pp. 215–216; Buchholz, 2015, p. 1786).

DSM-5 Diagnostic Criteria: Gender Dysphoria in Adolescents and Adults 302.85 (F64.1)

A. A marked incongruence between one's experienced/expressed gender and assigned gender, of at least 6 months' duration, as manifested by at least two of the following:
1. A marked incongruence between one's experienced/expressed gender and primary and/or secondary sex characteristics (or in young adolescents, the anticipated secondary sex characteristics).
2. A strong desire to be rid of one's primary and/or secondary sex characteristics because of a marked incongruence with one's experienced/expressed gender (or in young adolescents, a desire to prevent the development of the anticipated secondary sex characteristics).
3. A strong desire for the primary and/or secondary sex characteristics of the other gender.
4. A strong desire to be of the other gender (or some alternative gender different from one's assigned gender).
5. A strong desire to be treated as the other gender (or some alternative gender different from one's assigned gender).
6. A strong conviction that one has the typical feelings and reactions of the other gender (or some alternative gender different from one's assigned gender).
B. The condition is associated with clinically significant distress or impairment in social, occupational, or other important areas of functioning.

Treatments for Gender Dysphoria

In this appendix, we provide additional details about psychosocial, pharmacologic, surgical, and other treatments for gender dysphoria (GD).

Psychotherapy

The emphasis of psychotherapy for this population today is on "affirming a unique transgender identity," rather than focusing on gender transition (Institute of Medicine, 2011, p. 52). Mental health professionals can also help patients presenting with GD navigate the process of coming out to family, friends, and peers; treat comorbid mental health conditions;[1] weigh options related to gender identity, gender expression, and transition-related treatment interventions; and conduct assessments, make referrals, and guide preparation for and provide support through the transition-related treatment process (WPATH, 2011, pp. 22–26). Referral from a mental health professional is necessary under the standards of care for those seeking breast/chest or genital surgeries, and the latter also requires confirmation from an independent mental health provider (WPATH, 2011, p. 27). Mental health providers may also serve an important role on behalf of their patients by providing education and advocacy within the community and supporting changes to identity documents (WPATH, 2011, p. 31).

Of note, treatment aimed at changing one's gender identity to align with the sex assigned at birth has proven unsuccessful and is no longer considered ethical care; mental health providers who are unwilling or unable to provide appropriate care should refer patients to a provider who is (WPATH, 2011, p. 32).

Hormone Therapy

Hormone therapy is necessary for many individuals with GD (WPATH, 2011, p. 33). It has two major goals: (1) reduce naturally occurring hormones to minimize secondary sex characteristics and (2) maximize desired feminization/masculinization using the principles and medications used for hormone replacement in non-transgender patients who do not produce enough hormones, such as women who have had hyster-

---

[1] Co-occurring mental health conditions could range from anxiety and depression, which are common among the transgender population, to more severe and rare illnesses, such as schizophrenia or bipolar disorder.

ectomies or men with low testosterone (WPATH, 2011, p. 33; Hembree et al., 2009). As with most medications, there are risks, which may increase in the presence of some health conditions or behaviors (such as smoking); these should be evaluated and managed (Hembree et al., 2009).

For those transitioning from female to male, hormone therapy should lead to "deepened voice, clitoral enlargement (variable, 3–8 cm), growth in facial and body hair, cessation of menses, atrophy of breast tissue, increased libido, and increased percentage of body fat." For those transitioning from male to female, hormone therapy should lead to "breast growth (variable), decreased libido and erections, decreased testicular size, and increased percentage of body fat" (WPATH, 2011, p. 36). The timeline for these and other physical changes varies by individual; expected onset is within months, and maximum expected effect (such as body fat and muscle mass changes) is generally achieved in three or more years. Feminizing hormone therapy typically involves both estrogen and antiandrogens.[2] Masculinizing hormone therapy consists primarily of testosterone, which is available in oral, transdermal, parenteral (intravenous/intramuscular), buccal (cheek), and implantable administrations; brief use of progestin can help stop menstrual periods early in treatment (WPATH, 2011, p. 49). Detailed clinical practice guidelines are available from the Endocrine Society (Hembree et al., 2009).

Gender Transition–Related Surgery

As noted, gender transition–related surgery (also called sex reassignment surgery or gender-confirming surgery) is necessary for some transgender patients. Under the standards of care, mental health professionals must refer patients for surgery; in addition, criteria for both breast/chest and genital surgery include persistent and well-documented GD, the capacity to make informed decisions and to consent, and for other mental or general health concerns to be reasonably well controlled if present (WPATH, 2011, p. 59). Hormone therapy is not a prerequisite for breast/chest (also called "top") surgery, though it is recommended for 12–24 months for male-to-female patients to achieve optimal results (Hembree et al., 2009).

For genital (also called "bottom") surgery, 12 continuous months of hormone therapy are required prior to oophorectomy or orchiectomy (surgical removal of ovaries or testicles), unless contraindicated; health record documentation of "12 continuous months of living in a gender role that is congruent with their gender identity . . . consistently, on a day-to-day basis and across all settings of life" is also required for metoidioplasty (surgical relocation of an enlarged clitoris), phalloplasty (surgical creation of a penis), or vaginoplasty (surgical creation of a vagina; WPATH, 2011,

---

[2]   Transdermal rather than oral estrogen is recommended. Common antiandrogens include spironolactone (an antihypertensive agent that requires electrolyte monitoring); cyproterone acetate (not approved in the United States); GnRH agonists, such as gosrelin, buserelin, or triptorelin (available as injectables or implants); and 5-alpha reductase inhibitors, such as finasteride and dutasteride (WPATH, 2011, p. 48).

pp. 60–61). Mastectomy is often the only surgery undertaken by the female-to-male population; for those who do undergo genital surgery, phalloplasty is relatively uncommon, as it often requires multiple procedures and has frequent complications (WPATH, 2011, pp. 63–64). Surgeons should work closely with patients and other care providers, if needed, to ensure that the advantages, disadvantages, and risks of various treatments and procedures are well understood.

Other Treatments

Aside from breast/chest and genital surgery, other surgical interventions may include liposuction, lipofilling, and various aesthetic procedures. For male-to-female patients, these may include "facial feminization surgery, voice surgery, thyroid cartilage reduction, gluteal augmentation (implants/lipofilling), [and] hair reconstruction"; female-to-male patients may seek pectoral implants (WPATH, 2011, pp. 57–58). There is ongoing debate regarding whether these and other transition-related treatments are "medically necessary" (and therefore covered by insurance). For example, in some circumstances, facial hair removal for male-to-female patients may constitute necessary transition-related treatment: One study found that those who have undergone the procedure were "less likely to experience harassment in public spaces," and harassment can "have a negative impact on the success of a person's treatment for gender dysphoria" (Herman, 2013b, p. 19). In addition, voice and communication therapy to develop vocal characteristics and nonverbal communication patterns congruent with gender identity may prevent "vocal misuse and long-term vocal damage" (WPATH, 2011, pp. 52–54).

Review of Accession, Retention, and Separation Regulations

Directive	Date	Department
Air Force Instruction 36-2002, *Regular Air Force and Special Category Accessions*	4/7/1999, revised 6/2/2014	Air Force
Air Force Instruction Guidance Memorandum AFI48-123_AFGM2015-01, "Guidance Memorandum: AFI 48-123, *Medical Examinations and Standards*"	8/27/2015	Air Force
Air Force Instruction Guidance Memorandum 48-123_AFGM4, "Air Force Guidance Memorandum to AFI 48-123, *Medical Examinations and Standards*"	1/29/2013	Air Force
Air Force Recruiting Service Instruction 36-2001, *Recruiting Procedures for the Air Force*	8/1/2012	Air Force
Air Force Instruction 41-210, *TRICARE Operations and Patient Administration Functions*	6/6/2012	Air Force
U.S. Army Recruiting Command, *Pocket Recruiter Guide*	7/1/2013	Army
Army Regulation 635-40, *Physical Evaluation for Retention, Retirement, or Separation*	3/20/2012	Army
Army Regulation 601-280, *Army Retention Program*	9/15/2011	Army
Army Regulation 40-501, *Standards of Medical Fitness*	8/4/2011	Army
Army Regulation 40-66, *Medical Record Administration and Healthcare Documentation*	1/4/2010	Army
Army Regulation 635-200, *Active Duty Enlisted Administrative Separations*	9/6/2011	Army
Army Regulation 601-210, *Active and Reserve Components Enlistment Program*	3/12/2013	Army
DoDI 6130.03, *Medical Standards for Appointment, Enlistment, or Induction in the Military Services*	4/28/2010, revised 9/13/11	DoD
DoDI 1332.18, *Disability Evaluation System (DES)*	8/5/2014	DoD
Office of the Under Secretary of Defense for Personnel and Readiness, *Disability Evaluation System (DES) Pilot Operations Manual*	12/2008	DoD

Directive	Date	Department
Marine Corps Order 1040.31, *Enlisted Retention and Career Development Program*	9/8/2010	Marine Corps
Marine Corps Order 6110.3, *Marine Corps Body Composition and Military Appearance Program*	8/8/2008	Marine Corps
Marine Administrative Message 064/11, "Amplification to Testing Accession Standards for the Purpose of Application to Marine Office Commissioning Programs"	1/26/2011	Marine Corps
Navy Military Personnel Manual 1306-964, "Recruiting Duty"	5/9/2014	Navy
Navy Medicine Manual P-117, *Manual of the Medical Department*, Chapter 15, Article 15-31, "Waivers of Physical Standards"	5/3/2012	Navy and Marine Corps

References

American Psychiatric Association, *Diagnostic and Statistical Manual of Mental Disorders (DSM-III)*, 3rd ed., Arlington, Va., 1980.

———, *Diagnostic and Statistical Manual of Mental Disorders (DSM-III-R)*, 3rd ed., revised, Arlington, Va., 1987.

———, *Diagnostic and Statistical Manual of Mental Disorders (DSM-IV)*, 4th ed., revised, Arlington, Va., 1994.

———, *Diagnostic and Statistical Manual of Mental Disorders (DSM-5)*, 5th ed., Arlington, Va., 2013a.

———, "Gender Dysphoria," fact sheet, 2013b. As of January 5, 2016:
http://www.dsm5.org/documents/gender%20dysphoria%20fact%20sheet.pdf

APA—*See* American Psychiatric Association.

Army Regulation 40-501, *Standards of Medical Fitness*, December 14, 2007, revised August 4, 2011.

Army Regulation 600-8-101, *Personnel Processing (In-, Out-, Soldier Readiness, and Deployment Cycle)*, February 19, 2015.

Bakker, A., P. J. van Kesteren, L. J. Gooren, and P. D. Bezemer, "The Prevalence of Transsexualism in the Netherlands," *Acta Psychiatrica Scandinavica*, Vol. 87, No. 4, April 1993, pp. 237–238.

Belkin, Aaron, "Caring for Our Transgender Troops—The Negligible Cost of Transition-Related Care," *New England Journal of Medicine*, Vol. 373, No. 12, September 17, 2015, pp. 1089–1092.

Belkin, Aaron, and Jason McNichol, "Homosexual Personnel Policy in the Canadian Forces: Did Lifting the Gay Ban Undermine Military Performance?" *International Journal*, Vol. 56, No. 1, Winter 2000–2001, pp. 73–88.

Blakely, Katherine, and Don J. Jansen, *Post-Traumatic Stress Disorder and Other Mental Health Problems in the Military: Oversight Issues for Congress*, Washington, D.C.: Congressional Research Service, August 8, 2013.

Blosnich, John R., Adam J. Gordon, and Michael J. Fine, "Associations of Sexual and Gender Minority Status with Health Indicators, Health Risk Factors, and Social Stressors in a National Sample of Young Adults with Military Experience," *Annals of Epidemiology*, Vol. 25, No. 9, September 2015, pp. 661–667.

British Army, "Diversity," web page, undated. As of January 4, 2016:
http://www.army.mod.UK/join/38473.aspx

Brown, David, "Amputations and Genital Injuries Increase Sharply Among Soldiers in Afghanistan," *Washington Post*, May 4, 2011. As of January 5, 2016: https://www.washingtonpost.com/national/amputations-and-genital-injuries-increase-sharply-among-soldiers-in-afghanistan/2011/02/25/ABX0TqN_story.html

Brown, George R., "Transsexuals in the Military: Flight into Hypermasculinity," *Archives of Sexual Behavior*, Vol. 17, No. 6, December 1988, pp. 527–537.

Buchholz, Laura, "Transgender Care Moves into the Mainstream," *Journal of the American Medical Association*, Vol. 314, No. 17, November 3, 2015, pp. 1785–1787.

California Department of Health Services, *California Lesbian, Gay, Bisexual, and Transgender Tobacco Survey 2004*, San Francisco, Calif., 2004.

Canadian Armed Forces, Military Personnel Instruction 01/11, "Management of Transsexual Members," 2012.

Conron, Kerith, Gunner Scott, Grace Sterling Stowell, and Stewart J. Landers, "Transgender Health in Massachusetts: Results from a Household Probability Sample of Results," *American Journal of Public Health*, Vol. 102, No. 1, January 2012, pp. 118–122.

Cox, Matthew, "Army Has 50,000 Active Soldiers Who Can't Deploy, Top NCO Says," *Military.com*, November 25, 2015. As of March 16, 2016: http://www.military.com/daily-news/2015/11/25/army-has-50000-active-soldiers-who-cant-deploy-top-nco-says.html

De Cuypere, G., M. Van Hemelrijck, A. Michel, B. Carael, G. Heylens, R. Rubens, P. Hoebeke, and S. Monstrey, "Prevalence and Demography of Transsexualism in Belgium," *European Psychiatry*, Vol. 22, No. 3, 2007, pp. 137–141.

Defense Health Agency, TRICARE Management Activity, *Evaluation of the TRICARE Program: Access, Cost, and Quality, Fiscal Year 2015*, 2015. As of January 5, 2016: http://www.health.mil/Military-Health-Topics/Access-Cost-Quality-and-Safety/Health-Care-Program-Evaluation/Annual-Evaluation-of-the-TRICARE-Program

DoD—*See* U.S. Department of Defense.

Eklund, P. L., L. J. Gooren, and P. D. Bezemer, "Prevalence of Transsexualism in the Netherlands," *British Journal of Psychiatry*, Vol. 152, No. 5, May 1988, pp. 638–640.

Elders, Joycelyn, Alan M. Steinman, George R. Brown, Eli Coleman, and Thomas A. Kolditz, *Report of the Transgender Military Service Commission*, Santa Barbara, Calif.: Palm Center, March 2014.

Ender, Morten G., David E. Rohall, and Michael D. Matthews, "Cadet and Civilian Undergraduate Attitudes Toward Transgender People: A Research Note," *Armed Forces and Society*, Vol. 42, No. 2, April 2016, pp. 427–435.

Flores, Andrew R., "Attitudes Toward Transgender Rights: Perceived Knowledge and Secondary Interpersonal Contact," *Politics, Groups, and Identities*, Vol. 3, No. 3, 2015.

Frank, Nathaniel, *Gays in Foreign Militaries 2010: A Global Primer*, Santa Barbara, Calif.: Palm Center, 2010.

Gates, Gary J., *How Many People Are Lesbian, Gay, Bisexual, and Transgender?* Los Angeles, Calif.: Williams Institute, University of California, Los Angeles, School of Law, April 2011.

Gates, Gary J., and Jody L. Herman, "Transgender Military Service in the United States," Los Angeles, Calif.: Williams Institute, University of California, Los Angeles, School of Law, May 2014.

Gijs, Luk, and Anne Brewaeys, "Surgical Treatment of Gender Dysphoria in Adults and Adolescents: Recent Developments, Effectiveness, and Challenges," *Annual Review of Sex Research*, Vol. 18, No. 1, 2007, pp. 178–224.

Gould, Elise, *A Decade of Declines in Employer-Sponsored Health Insurance Coverage*, Washington, D.C.: Economic Policy Institute, February 2012. As of January 5, 2016: http://www.epi.org/publication/bp337-employer-sponsored-health-insurance

Grant, Jaime M., Lisa A. Mottet, and Justin Tanis, with Jack Harrison, Jody L. Herman, and Mara Keisling, *Injustice at Every Turn: A Report of the National Transgender Discrimination Survey*, Washington, D.C.: National Center for Transgender Equality and National Gay and Lesbian Task Force, 2011.

Harrell, Margaret C., Laura Werber, Peter Schirmer, Bryan W. Hallmark, Jennifer Kavanagh, Daniel Gershwin, and Paul S. Steinberg, *Assessing the Assignment Policy for Army Women*, Santa Monica, Calif.: RAND Corporation, MG-590-1-OSD, 2007. As of March 17, 2016: http://www.rand.org/pubs/monographs/MG590-1.html

Harris, Benjamin Cerf, *Likely Transgender Individuals in the U.S. Federal Administrative Records and the 2010 Census*, Washington, D.C.: U.S. Census Bureau, May 4, 2015.

Hembree, Wylie C., Peggy Cohen-Kettenis, Henriette A. Delemarre–van de Waal, Louis J. Gooren, Walter J. Meyer III, Norman P. Spack, Vin Tangpricha, and Victor M. Montori, "Endocrine Treatment of Transsexual Persons: An Endocrine Society Clinical Practice Guideline," *Journal of Clinical Endocrinology and Metabolism*, Vol. 94, No. 9, September 2009, pp. 3132–3154.

Herman, Jody L., *The Cost of Employment and Housing Discrimination Against Transgender Residents of New York*, Los Angeles, Calif.: Williams Institute, University of California, Los Angeles, School of Law, April 2013a.

———, *Costs and Benefits of Providing Transition-Related Health Care Coverage in Employee Health Benefits Plans: Findings from a Survey of Employers*, Los Angeles, Calif.: Williams Institute, University of California, Los Angeles, School of Law, September 2013b.

Hoenig, J., and J. C. Kenna, "The Prevalence of Transsexualism in England and Wales," *British Journal of Psychiatry*, Vol. 124, No. 579, 1974, pp. 181–190.

Hoge, Charles W., Jennifer Auchterlonie, and Charles S. Millike, "Mental Health Problems, Use of Mental Health Services, and Attrition from Military Service After Returning from Deployment to Iraq or Afghanistan," *Journal of the American Medical Association*, Vol. 295, No. 9, March 1, 2006, pp. 1023–1032.

Horton, Mary Ann, "The Incidence and Prevalence of SRS Among US Residents," paper presented at the Out and Equal Workplace Summit, September 12, 2008. As of January 5, 2016: http://www.tgender.net/taw/thb/THBPrevalence-OE2008.pdf

Institute of Medicine, *The Health of Lesbian, Gay, Bisexual, and Transgender People: Building a Foundation for Better Understanding*, Washington, D.C.: National Academies Press, 2011.

Intersex Society of North America, "What Is Intersex?" web page, undated. As of January 5, 2016: http://www.isna.org/faq/what_is_intersex

Kates, Jen, Usha Ranji, Adara Beamesderfer, Alina Salganicoff, and Lindsey Dawson, *Health and Access to Care and Coverage for Lesbian, Gay, Bisexual, and Transgender Individuals in the U.S.*, Menlo Park, Calif.: Henry J. Kaiser Family Foundation, July 2015.

Kauth, Michael R., Jillian C. Shipherd, Jan Lindsay, John R. Blosnich, George R. Brown, and Kenneth T. Jones, "Access to Care for Transgender Veterans in the Veterans Health Administration: 2006–2013," *American Journal of Public Health*, Vol. 104, No. S4, September 2014, pp. S532–S534.

Keen, Lisa, "Mass. Ranks Sixth for LGBT-Friendly Laws, Study Says," *Boston Globe*, May 28, 2015. As of March 17, 2016:
https://www.bostonglobe.com/news/politics/2015/05/27/mass-ranks-sixth-for-lgbt-friendly-laws-study-says/sBX5TpZdNeusUo7Iuqs2qN/story.html

Lambda Legal, "Professional Organization Statements Supporting Transgender People in Health Care," last updated June 8, 2012. As of January 4, 2016:
http://www.lambdalegal.org/sites/default/files/publications/downloads/fs_professional-org-statements-supporting-trans-health_1.pdf

McKibben, Jodi B. A., Carol S. Fullerton, Christine L. Gray, Ronald C. Kessler, Murray B. Stein, and Robert J. Ursano, "Mental Health Service Utilization in the U.S. Army," *Psychiatric Services*, Vol. 64, No. 4, April 2013, pp. 347–353.

Milhiser, Mark R., "Transgender Service: The Next Social Domino for the Army," *Military Law Review*, Vol. 220, Summer 2014, pp. 191–217.

Navy Medical Policy 07-009, *Deployment Medical Readiness*, April 6, 2007.

Norton, Aaron T., and Gregory M. Herek, "Heterosexuals' Attitudes Toward Transgender People: Findings from a National Probability Sample of U.S. Adults," *Sex Roles*, Vol. 68, No. 11, June 2013, pp. 738–753.

Office of the Assistant Secretary of Defense for Health Affairs, "Policy for Cosmetic Surgery Procedures in the Military Health System," Health Affairs Policy 05-020, October 25, 2005.

———, "Clinical Practice Guidance for Deployment-Limiting Mental Disorders and Psychotropic Medications," memorandum, October 7, 2013.

Office of Personnel Management, *Addressing Sexual Orientation and Gender Identity Discrimination in Federal Civilian Employment*, Washington, D.C., June 2015.

Okros, Alan, and Denise Scott, "Gender Identity in the Canadian Forces," *Armed Forces and Society*, Vol. 41, No. 2, April 2015, pp. 243–256.

Padula, William V., Shiona Heru, and Jonathan D. Campbell, "Societal Implications of Health Insurance Coverage for Medically Necessary Services in the U.S. Transgender Population: A Cost-Effectiveness Analysis," *Journal of General Internal Medicine*, October 19, 2015.

Parco, James E., David A. Levy, and Sarah R. Spears, "Transgender Military Personnel in the Post-DADT Repeal Era: A Phenomenological Study," *Armed Forces and Society*, Vol. 41, No. 2, 2015, pp. 221–242.

Polchar, Joshua, Tim Sweijs, Phillip Marten, and Jan Gladega, *LGBT Military Personnel: A Strategic Vision for Inclusion*, The Hague, Netherlands: The Hague Centre for Strategic Studies, 2014.

Pollock, Gale S., and Shannon Minter, *Report of the Planning Commission on Transgender Military Service*, Santa Barbara, Calif.: Palm Center, August 2014.

RAND National Defense Research Institute, *Sexual Orientation and U.S. Military Personnel Policy: An Update of RAND's 1993 Study*, Santa Monica, Calif.: RAND Corporation, MG-1056-OSD, 2010. As of March 17, 2016:
http://www.rand.org/pubs/monographs/MG1056.html

Rashid, Mamoon, and Muhammad Sarmad Tamimy, "Phalloplasty: The Dream and the Reality," *Indian Journal of Plastic Surgery*, Vol. 46, No. 2, May 2013, pp. 283–293.

Reed, Bernard, Stephenne Rhodes, Pietà Schofiled, and Kevan Wylie, *Gender Variance in the UK: Prevalence, Incidence, Growth and Geographic Distribution*, Surrey, UK: Gender Identity Research and Education Society, June 2009.

Roller, Cyndi Gale, Carol Sedlak, and Claire Burke Draucker, "Navigating the System: How Transgender Individuals Engage in Health Care Services," *Journal of Nursing Scholarship*, Vol. 47, No. 5, September 2015, pp. 417–424.

Ross, Allison, "The Invisible Army: Why the Military Needs to Rescind its Ban on Transgender Service Members," *Southern California Interdisciplinary Law Journal*, Vol. 23, No. 1, 2014, pp. 185–216.

Rostker, Bernard D., Scott A. Harris, James P. Kahan, Erik J. Frinking, C. Neil Fulcher, Lawrence M. Hanser, Paul Koegel, John D. Winkler, Brent A. Boultinghouse, Joanna Heilbrunn, Janet Lever, Robert J. MacCoun, Peter Tiemeyer, Gail L. Zellman, Sandra H. Berry, Jennifer Hawes-Dawson, Samantha Ravich, Steven L. Schlossman, Timothy Haggarty, Tanjam Jacobson, Ancella Livers, Sherie Mershon, Andrew Cornell, Mark A. Schuster, David E. Kanouse, Raynard Kington, Mark Litwin, Conrad Peter Schmidt, Carl H. Builder, Peter Jacobson, Stephen A. Saltzburg, Roger Allen Brown, William Fedorochko, Marilyn Fisher Freemon, John F. Peterson, and James A. Dewar, *Sexual Orientation and U.S. Military Personnel Policy: Options and Assessment*, Santa Monica, Calif.: RAND Corporation, MR-323-OSD, 1993. As of March 17, 2016: http://www.rand.org/pubs/monograph_reports/MR323.html

Royal Australian Air Force, *Air Force Diversity Handbook: Transitioning Gender in Air Force*, July 2015.

Schaefer, Agnes Gereben, Jennie W. Wenger, Jennifer Kavanagh, Jonathan P. Wong, Gillian S. Oak, Thomas E. Trail, and Todd Nichols, *Implications of Integrating Women into the Marine Corps Infantry*, Santa Monica, Calif.: RAND Corporation, RR-1103-USMC, 2015. As of March 17, 2016: http://www.rand.org/pubs/research_reports/RR1103.html

Sonier, Julie, Brett Fried, Caroline Au-Yeung, and Breanna Auringer, *State-Level Trends in Employer-Sponsored Health Insurance, A State-by-State Analysis*, Minneapolis, Minn.: State Health Access Data Center and Robert Wood Johnson Foundation, April 2013.

Speckhard, Anne, and Reuven Paz, "Transgender Service in the Israeli Defense Forces: A Polar Opposite Stance to the U.S. Military Policy of Barring Transgender Soldiers from Service," unpublished research paper, 2014. As of January 4, 2016: http://www.researchgate.net/publication/280093066

State of California, Department of Insurance, "Economic Impact Assessment: Gender Nondiscrimination in Health Insurance," Regulation File Number: REG-2011-00023, April 13, 2012. As of January 5, 2016: http://transgenderlawcenter.org/wp-content/uploads/2013/04/Economic-Impact-Assessment-Gender-Nondiscrimination-In-Health-Insurance.pdf

Szayna, Thomas S., Eric V. Larson, Angela O'Mahony, Sean Robson, Agnes Gereben Schaefer, Miriam Matthews, J. Michael Polich, Lynsay Ayer, Derek Eaton, William Marcellino, Lisa Miyashiro, Marek Posard, James Syme, Zev Winkelman, Cameron Wright, Megan Zander-Cotugno, and William Welser, *Considerations for Integrating Women into Closed Occupations in the U.S. Special Operations Forces*, Santa Monica, Calif.: RAND Corporation, RR-1058-USSOCOM, 2015. As of March 17, 2016: http://www.rand.org/pubs/research_reports/RR1058.html

Tan, Michelle, "SMA Calls for Bonus Money for Soldiers on Deployment, at NTC," *Army Times*, November 1, 2015. As of March 16, 2016: http://www.armytimes.com/story/military/benefits/pay/allowances/2015/11/01/sma-calls-bonus-money-soldiers-deployment-ntc/74821828

Tsoi, W. F., "The Prevalence of Transsexualism in Singapore," *Acta Psychiatrica Scandinavica*, Vol. 78, No. 4, 1988, pp. 501–504.

UK Ministry of Defence, "Policy for the Recruitment and Management of Transsexual Personnel in the Armed Forces," January 2009.

UnitedHealthcare, "Gender Dysphoria (Gender Identity Disorder) Treatment," Coverage Determination Guideline CDG.011.05, effective October 1, 2015. As of January 5, 2016:
https://www.unitedhealthcareonline.com/ccmcontent/ProviderII/UHC/en-US/Assets/ProviderStaticFiles/ProviderStaticFilesPdf/Tools%20and%20Resources/Policies%20and%20Protocols/Medical%20Policies/Medical%20Policies/Gender_Identity_Disorder_CD.pdf

U.S. Central Command, "PPG-TAB A: Amplification of the Minimal Standards of Fitness for Deployment to the CENTCOM AOR; to Accompany MOD ELEVEN to USCENTCOM Individual Protection and Individual/Unit Deployment Policy," December 2, 2013. As of March 17, 2016:
http://www.tam.usace.army.mil/Portals/53/docs/UDC/medical-disqualifiers.pdf

U.S. Department of Defense, *2014 Demographics: Profile of the Military Community*, Washington, D.C., 2014. As of January 5, 2016:
http://download.militaryonesource.mil/12038/MOS/Reports/2014-Demographics-Report.pdf

———, "DoD Announces Recruiting and Retention Numbers for Fiscal 2015, Through November 2014," press release, No. NR-001-15, January 6, 2015a. As of January 4, 2016:
http://www.defense.gov/News/News-Releases/News-Release-View/Article/605335

———, "Statement by Secretary of Defense Ash Carter on DoD Transgender Policy," press release, No. NR-272-15, July 15, 2015b. As of March 16, 2016:
http://www.defense.gov/News/News-Releases/News-Release-View/Article/612778

U.S. Department of Defense Instruction 1332.14, *Enlisted Administrative Separations*, January 27, 2014, incorporating change 1, December 4, 2014.

U.S. Department of Defense Instruction 1332.18, *Disability Evaluation System (DES)*, August 5, 2014.

U.S. Department of Defense Instruction 1332.30, *Separation of Regular and Reserve Commissioned Officers*, November 25, 2013.

U.S. Department of Defense Instruction 1332.38, *Physical Disability Evaluation*, November 14, 1996, incorporating change 1, July 10, 2006.

U.S. Department of Defense Instruction 6130.03, *Medical Standards for Appointment, Enlistment, or Induction in the Military Services*, April 28, 2010, incorporating change 1, September 13, 2011.

U.S. Government Accountability Office, *Personnel and Cost Data Associated with Implementing DOD's Homosexual Conduct Policy*, Washington, D.C., GAO-11-170. January 2011.

Van Kesteren, Paul J., Louis J. Gooren, and Jos A. Megens, "An Epidemiological and Demographic Study of Transsexuals in the Netherlands," *Archives of Sexual Behavior*, Vol. 25, No. 6, 1996, pp. 589–600.

Wålinder, Jan, "Transsexualism: Definition, Prevalence and Sex Distribution," *Acta Psychiatrica Scandinavica*, Vol. 43, No. S203, August 1968, pp. 255–257.

———, "Incidence and Sex Ratio of Transsexualism in Sweden," *British Journal of Psychiatry*, Vol. 119, No. 549, 1971, pp. 195–196.

Wallace, Duncan, "Trends in Traumatic Limb Amputation in Allied Forces in Iraq and Afghanistan," *Journal of Military and Veterans' Health*, Vol. 20, No. 2, April 2012.

Weitze, Cordula, and Susanne Osburg, "Transsexualism in Germany: Empirical Data on Epidemiology and Application of the German Transsexuals' Act During Its First Ten Years," *Archives of Sexual Behavior*, Vol. 25, No. 4, 1996, pp. 409–425.

Welsh, Ashley, "First U.S. Penis Transplants Planned to Help Wounded Vets," CBS News, December 7, 2015. As of January 5, 2016:
http://www.cbsnews.com/news/first-penis-transplants-planned-in-u-s-to-help-wounded-vets

Williams, Molly, and James Jezior, "Management of Combat-Related Urological Trauma in the Modern Era," *Nature Reviews Urology*, Vol. 10, No. 9, September 2013, pp. 504–512.

World Professional Association for Transgender Health, *Standards of Care for the Health of Transsexual, Transgender, and Gender Nonconforming People*, version 7, Elgin, Ill., 2011.

WPATH—*See* World Professional Association for Transgender Health.

Yerke, Adam F., and Valory Mitchell, "Transgender People in the Military: Don't Ask? Don't Tell? Don't Enlist!" *Journal of Homosexuality*, Vol. 60, Nos. 2–3, 2013, pp. 436–457.

Zitun, Yoav, "IDF to Support Transgender Recruits Throughout the Sex Change Process," *YNET News*, December 26, 2014. As of January 4, 2016:
http://www.ynetnews.com/articles/0,7340,L-4608141,00.html

Zucker, Kenneth J., Susan J. Bradley, Allison Owen-Anderson, Sarah J. Kibblewhite, and James M. Cantor, "Is Gender Identity Disorder in Adolescents Coming out of the Closet?" *Journal of Sex and Marital Therapy*, Vol. 34, No. 4, June 2008, pp. 287–290.

Zucker, Kenneth J., and Anne A. Lawrence, "Epidemiology of Gender Identity Disorder: Recommendations for the Standards of Care of the World Professional Association for Transgender Health," *International Journal of Transgenderism*, Vol. 11, No. 1, 2009, pp. 8–18.